Summer
READING LIST 2023

J.P.Morgan

Praise for *Money Machine*

"Weijian Shan has been at the top of the Asian private equity game for decades. But this is much more than just a gripping account of the vicissitudes of one deal. It is the story of the only foreign investor to take control of a major Chinese bank and throws light on China's gigantic life-or-death struggle to rescue its banking sector as it stood teetering on the edge of the abyss."

—Tim Clissold,
Author of *Mr. China*

"Shan Weijian is a master dealmaker who takes the reader on a white-knuckled ride on the roller coaster of China's reform era. It should be required reading in business schools and by anybody who enjoys an intriguing story of money making."

—James McGregor,
30 years in China as a journalist, businessman,
investor, consultant, and author of
No Ancient Wisdom, No Followers, and *One Billion Customers*

"The book shows private equity at its very best . . . taking a perilously close to insolvent institution with bad loans, cleaning it up through write downs and write offs, and powering it forward with good governance and innovative new products on the back of a booming Chinese economy. It's a great case study of how private equity can transform underperforming enterprises even in China... but it took many false starts by Shan to achieve the control necessary for the American firm to administer the medicine needed to transform the business into the trophy asset it became. The storied turnaround, orchestrated by Shan and his partners, was a tour de force that has helped burnish Shan's reputation as an imaginative and capable titan of private equity in Asia."

—David J. Teece,
Professor of The Graduate School at the
University of California, Berkeley, the faculty
director of the school's Tusher Initiative for the
Management of Intellectual Capital and
a prolific author of books and scholarly papers

Money Machine

Money Machine

A Trailblazing American
Venture in China

Weijian Shan

WILEY

Library of Congress Cataloging-in-Publication Data

Names: Shan, Weijian, 1953- author.
Title: Money machine : a trailblazing American venture in China / Weijian Shan.
Description: Hoboken, New Jersey : John Wiley & Sons, Inc., [2023] | Includes index.
Identifiers: LCCN 2022049588 (print) | LCCN 2022049589 (ebook) | ISBN 9781394161201 (hardback) | ISBN 9781394161225 (adobe pdf) | ISBN 9781394161218 (epub)
Subjects: LCSH: Shenzhen fa zhan yin hang. | Banks and banking—China—Case studies. | Banks and banking—Foreign ownership—Case studies. | Bank management—China—Case studies. | Private equity—United States. | Global Financial crisis, 2008–2009.
Classification: LCC HG3338.S54 S53 2023 (print) | LCC HG3338.S54 (ebook) | DDC 332.10951—dc23/eng/20221019
LC record available at https://lccn.loc.gov/2022049588
LC ebook record available at https://lccn.loc.gov/2022049589

Cover Design: Wiley
Cover Image: © Koiyip Lam/Getty Images
SKY10040147_121622

To our investors, to whom we owe our gratitude for their partnership and trust over the years.

Contents

Foreword

An Extraordinary Journey

During the Global Financial Crisis (GFC), banks around the world, including in America, Britain, and Europe, were supported by government cash injections and guarantees.

But there was one significant bank, deeply troubled before then, that worked through that time to strength and success, without *any* government funding or financial guarantees.

Where was this bank? In China! And who controlled the bank and guided it during this extraordinary recovery? A U.S.-based private equity firm!

How was all this possible? How was it possible that a U.S. firm took on leadership of a Chinese bank? How did a good-sized national bank, troubled and weak before the GFC, become strong, healthy, and successful, without a single dollar of government support? How did the team work with Chinese regulators to implement a series of unprecedented steps? How did the private-equity process proceed, in China, once the bank was so successful, to realize very substantial benefits for its investors, in U.S. dollars?

And there's more—all true, amazingly. A high-profile lawsuit filed in the United States, involving a Chinese bank and a Taiwanese-listed company, and of interest to the Chinese government at high levels. A board composed largely of Chinese businesspeople, with a U.S.-based major shareholder and an American chairman, making rapid change.

Weijian Shan tells this extraordinary success story, based on his own leadership role, in a fascinating insider's chronicle, including insights into the many challenges of business in China.

Shan led the complex process of establishing an American private equity firm as the lead shareholder. He led the bank through a uniquely challenging process of raising capital. And, once the bank was fully recovered to strength and profitability, Shan led another complex but highly successful process of the sale of the bank, finding a good home for its future, and rewarding investors handsomely.

I had the privilege of leading the management of the bank during most of that time, as its chairman and CEO. I was the only Western employee in the bank, the only one who did not speak Chinese. It was a very special experience. I am proud of my Chinese colleagues who worked diligently with me to take the strategic and tactical steps to turn the bank around, from weak and troubled to strong and successful. I still have many friends there.

Shan tells this extraordinary story from deep inside knowledge, with fascinating insights and perspectives. It's a great story for readers who want to understand more of how things can really work inside China's capital system, and for those who want to see how a private equity firm can add real value to a company, even in the face of exceptional challenges.

Frank Newman
Chairman and CEO, Shenzhen Development Bank (2005–2010)
Chairman and CEO, Bankers Trust (1996–1999)
Undersecretary/Deputy Secretary, United States Department of the Treasury (1993–1995)
October 31, 2021

A Note from the Author

This story involves several Chinese customs and systems that may not be familiar to Western readers. These involve names, Chinese currency, and the organization of the Chinese government. Here, I outline some important facts to help readers best understand the details of the story.

People in China present their names in the proper Chinese order: family name first, followed by their given name. For example, Mao, Zhou, and Chiang are surnames of, respectively, Mao Zedong, Zhou Enlai (China's former prime minister), and Chiang Kai-shek (former leader of China's Nationalist government). East Asian countries such as Japan, Korea, and Vietnam all present their names in this order. To my knowledge, Hungary is the only European country that uses this approach for names.

When I first arrived in the United States as a student in the 1980s, I wrote my name the Chinese way, family name first followed by the given name: Shan Weijian. Soon afterward, I learned that Americans put their given name first, followed by the family name; so they call the given name "first name" and the family name "last name." Needless to say, it was confusing to people that my first name is actually my last name. Americans usually greet each other by their given, or first, names. So, my new friends called me Shan. I was totally fine with it, because my given name, Weijian, is harder to pronounce or remember. Now English-speaking people around me call me Shan, and I always sign my letters in English with just *Shan*.

To avoid confusion, I later conformed to the American way of writing my name: Weijian Shan.

This book presents Chinese names following the East Asian order: family name first, followed by the given name. If a Chinese person has taken a Western name—not uncommon for Western-educated Chinese—I present their names in the Western way, such as Alex Zhang, Daniel Poon, Fred Hu, Louis Cheung, and Rachel Kwok. And for Westerners, names are presented as usual: David Bonderman, Frank Newman, or Oprah Winfrey.

When people in the same passage share a name, I use their full names for clarity. In China, there are hundreds of common surnames but millions of given names, each of which may be unique. This is just the opposite of the Anglo-Saxon world in which there are numerous family names but a very limited number of given names. My firm, PAG, has so many Davids that we have to refer to them by their initials or full names.

A final note about names: In mainland China, a woman keeps her own family name after marriage, so there is no such concept as a maiden name. When I was growing up in China, there were no wedding rings, although some young Chinese people today adopt Western ways. How do you tell if a middle-aged Chinese woman or man is married? You just have to ask.

Now, let's talk briefly about Chinese currency. The Chinese currency is called *renminbi* (RMB), which roughly translates to "the people's money." The unit of currency similar to the dollar is the *yuan*. In fact, the word *dollar* is translated in Chinese as *yuan*, so the U.S. dollar is *mei yuan* in Chinese (*mei* refers to the United States). The RMB is convertible into foreign currencies for the purposes of trade, services, and direct foreign investments. However, for investments in the securities markets, such as stocks and bonds, there are various restrictions. Therefore, at the time of this writing, and at the time the events of this book take place, the RMB was not fully convertible. The aim of the

Chinese government is to make its currency fully and freely convertible in the future.

Finally, the government system of China is that of a *unitary* state in which the central government has the ultimate power, and under it are provinces, cities, districts, and so forth. In contrast, the United States is a *federalist* state, a union of partially self-governing states under a federal government. The unitary state is the more common form of government—166 of the 193 United Nations members—have a unitary system of government. In China, there is another layer of governance. The central government, headed by the president, followed by the prime minister, is not the highest decision-making body.

The highest decision-making body in China is the political bureau, or the Politburo, of the Chinese Communist Party (the CCP). Therefore, the chairman or the general secretary of the CCP—not the prime minister or president—is the top leader (although in the past three decades, the party secretary has also been the president). Mao Zedong ruled China as the CCP chairman; he never took a position in the government and refused the title of president. After Mao's death, Deng Xiaoping was China's paramount leader, even though his position in the government had never been higher than that of vice premier. Within the CCP, Deng was not the chairman or general secretary, but he was the chairman of CCP's central military commission. Because the CCP is the ruling party, party secretaries at different levels within the government have more seniority than the highest government administrators who are not also party secretaries. For example, the party secretary of a province ranks higher than the governor of that province, and the party secretary of a city has a higher position than the mayor of that city.

Weijian Shan
August 2022

Part I

Courtship

Chapter 1

An Unexpected Question

From aboard the Star Ferry in Victoria Harbor, your view of Hong Kong Island builds, growing more dramatic as the boat makes its short trip across the water. From the Kowloon side, you can make out the bends in the Hong Kong waterfront, a mix of malls and parks, ferry landings and exhibition halls, all on flat land. But as the ferry nears its destination, the eye is drawn skyward, to the forests beyond the waterfront—actual forests of trees and fast-rising jungle, and forests of high-rises that wink and glitter in the shifting light, competing with one another as they reach for the sun. Depending on the time of year, the weather, and where you are in the harbor, those buildings really do seem to scrape against the clouds.

Coming into port with the Star Ferry, you land at Central Pier, where the choice lies at your feet. You will head into one forest or another—the natural or the man–made—on an island that boasts, among other things, one of the densest populations and some of the costliest real estate anywhere on Earth.

On this particular morning in 2002, Hong Kong was sweltering. The spring fog that shrouds Victoria Peak had burned off under a blazing sun. Warm as it was outside, I knew that in those skyscrapers the offices would feel cold. Hong Kong natives have a habit of blasting their air conditioning as if to overcompensate for the heat and humidity outside.

It was Monday, April 1, and I was on my way to my company's offices, which were located by the harborfront at One International Financial Centre in Central, a 688-foot office tower with 39 stories, connected with a sprawling shopping mall below and the Four Seasons Hotel beside it. Once I reached our conference room on the 20th floor, I really was chilled by the air conditioning. Waiting for a meeting to begin, I took in the blue sky dotted with a few white clouds. In the distance, looming behind the vast expanse of numerous low- and high-rises in the city of Kowloon was Lion Rock, whose shape resembles a crouching lion, in the middle of a range of lush green mountains.

It was a lovely day, and I thought for a moment that it might augur well for the city's air quality; for much of the winter, the city had suffered under a haze of smog. But April brings summer weather to Hong Kong, and with it comes a shift in the winds. On this morning you could already sense the change. The haze was gone, and fresh sea air from the south had taken its place.

Perhaps the fine weather would also be a harbinger of things to come in that conference room.

My firm was Newbridge Capital, a private equity investment partnership established in 1994 by Richard C. Blum and David Bonderman. Blum was also the founder and chairman of Richard C. Blum & Associates (later renamed Blum Capital Partners), and Bonderman was the chairman of Texas Pacific Group (later renamed TPG).

My visitor on this April morning was Alex Zhang, a partner at Dorsey & Whitney, an American law firm headquartered in Minneapolis, with a sizable office in Hong Kong. He was a friend of many years, a veteran lawyer in a firm whose partners boasted the likes of former U.S. vice president Walter Mondale. Zhang was handsome, about six feet tall, and to me he looked like Yao Ming, the NBA basketball star, albeit a foot and a half shorter. He was unmarried and 42, but looked younger, and it had occurred to me that he made for a highly eligible bachelor. He graduated from the Institute of International Relations in Beijing and went on to receive law degrees from the China University of Political Science and Law and from the University of Minnesota before he joined Dorsey & Whitney about a dozen years earlier.

Sitting there in our conference room, just the two of us, Zhang came right to the point. "Shan," he asked, "do you think Newbridge Capital is interested in buying a Chinese bank?" The question was unexpected. I felt a shot of workplace adrenaline. He was smiling. I may have smiled in return.

★ ★ ★

This is a book about a bank deal, about how it happened, and the reverberations it sent across China and the world of high finance. More than that, it is the story of a deal that changed everything, not only for the people and institutions involved, but for many who, to this day, may have never heard of Newbridge Capital, the Shenzhen Development Bank, or the people who brought the deal to fruition. It is also very much a story about change in China, and how American private equity investors, at least some of them, have thrived there.

Countless pages have been written and speeches given about "the New China." Depending on one's frame of reference, China was "new" in 1992, again in 2002, in 2012, and so on. At all those moments the analyses and assessments of change have come from all corners. Those of us who have lived and worked in this "new" and growing China perhaps didn't notice the changes the way that outsiders did—much the same way that a parent may not notice a child's growth spurt the way a family friend might. Over the past few decades, China has seemed "new" to anyone who has not been in the country the previous year or so. Today, a first-time visitor will often be told, *Oh, you should have been here five years ago—and seen how things were.* Such is the nature and pace of change in the country.

But in any historical, tectonic shift, once in a while there are moments that really do shine a light on something more profound—events that with the benefits of reflection and hindsight contribute to a greater understanding of the turns of history.

Certainly I could not have imagined that Zhang's April visit to our office might carry such significance. He was one more intermediary, bringing one more idea, and one can only imagine how many individuals were making a financial pitch in Hong Kong on any given day, trying to sell a concept or a company, or just a new way of doing things. How many people must have been doing exactly that along the Hong Kong harborfront on that very same morning? For all I know, others could have been pitching their ideas in the same skyscraper where Zhang had come to make his case.

And yet, looking back, it is hard to dispute the fact that Zhang's question would change the trajectory of things—for a company, for a major bank, for my colleagues, and for China as well.

"Yes," I told him on that April morning. "We are interested to know more."

★ ★ ★

His question was not as abrupt as it might have sounded. Newbridge Capital was well known in Asia's financial markets. We had received a good deal of press just a few years earlier, after acquiring control of Korea First Bank (KFB) from the government of South Korea. For a time, KFB had been the largest bank in Korea, but during the 1997–1998 Asian Financial Crisis the bank had failed and been nationalized. We had come in soon after, and the ensuing transaction had taken months to negotiate and consummate. Our negotiations with the Korean government had read in the press like some never-ending soap opera, dragging on until the deal was done in January 2000. But the story had ended well for all concerned. (Those interested in learning more should read my book *Money Games: The Inside Story of How American Dealmakers Saved Korea's Most Iconic Bank*.) The deal had received enough good publicity that Newbridge was perceived in financial circles as a true turnaround specialist and a credible buyer of banks, especially the troubled ones. We seemed to know what to do with them.

So for Zhang to come to Newbridge with that question made sense. As for the question itself, of *buying a Chinese bank*—well, it had never been done. Foreign investors had taken small stakes in Chinese banks, but no institution, domestic or foreign, had ever bought outright control. That said, in those years, nearly every day brought some development in the Chinese economy that had never happened before.

I listened to Zhang, expressed our general interest, and then I chose my words carefully. There would be much to consider, I knew, but one point was paramount in my mind. "It will depend," I began, "on whether or not we'll be able to control the bank. We wouldn't be interested in being a passive investor."

Newbridge had managed to turn around KFB because we acquired full control. That had allowed us to appoint our own management team, install our own risk-management system, and create a credit culture—a set of beliefs and practices based on the credit and financial conditions of the borrower—that ensured asset quality. Without such control, we would never have been able to turn it around, nor win credibility in the media and financial circles. Chinese banks were a whole other matter. They were notorious for high levels of bad loans. No sophisticated investor would want to take a ride as a passive shareholder with banks of poor-quality assets.

"The bank in question is Shenzhen Development Bank," Zhang said. "Do you know it?"

I knew this much about Shenzhen Development Bank (SDB): It was a nationwide bank based in Shenzhen, a booming city just across the border from Hong Kong. The city had become a kind of poster child for the explosive economic growth in southeastern China. It was also one of 13 joint-stock commercial banks in China. The largest Chinese banks were 100% owned and controlled by the government. Joint-stock banks had more diversified shareholding that included private ownership. We knew the model, but we did not know much about SDB itself.

Zhang filled in some blanks. SDB was controlled by the city government of Shenzhen, which was considering selling its controlling stake to a foreign investor. This much was not surprising. Since unleashing its economic reforms more than 20 years earlier, China had been moving toward a market economy, including the privatization, overhaul, and cleansing of state-owned enterprises—especially those in competitive industries. For this purpose, foreign direct investment had been strongly encouraged, not only for the capital but also for the expertise that came with it.

The city of Shenzhen had been China's first Special Economic Zone, chosen by Deng Xiaoping (1904–1997), China's paramount leader and architect of the market reforms. It was a grand experiment that had begun in 1978 to loosen the strictures of a centrally planned economy. In two decades, Shenzhen had morphed from a fishing village into a high-tech hub and a megalopolis of more than 10 million people. As such, it had adopted policies that were more market-friendly than those elsewhere in the country. Shenzhen was at the frontier of market-oriented reforms.

But for all the dizzying changes that had come to China, there was no precedent for foreign control of a Chinese national bank. Chinese regulations were generally designed to keep any single private investor from controlling a bank, while severely limiting foreign ownership. This was hardly a China-only phenomenon; banking regulations all over the world restrict foreign takeover of banks. And we knew that if there was to be a breakthrough, it could only happen in Shenzhen.

All of which we understood. And it all made sense. But to put our initial concerns mildly, buying a Chinese bank was not for the fainthearted. The market consensus was that China's banking system was technically insolvent and badly in need of wholesale reforms. The main issue was nonperforming loans (NPLs). These were loans with borrowers who had defaulted on their debt, or simply stopped making payments. Almost all Chinese banks were laden with NPLs, unable to stand on their own without government support.

How weak exactly was the banking sector? The numbers—to the extent they could be known—boggled the mind. The People's Bank of China (PBOC), China's central bank, estimated that in 2002, as much as 30% of the loans held by the country's four biggest banks—the Industrial and Commercial Bank of China, China Construction Bank, Bank of China, and Agricultural Bank of China—were nonperforming. According to Standard & Poor's, that 30% ratio was the figure *after* the banks had "significantly improved their credit and risk management controls in recent years." S&P painted a grim picture in one of its reports:

> *Standard & Poor's estimates that the country's banking sector, acting alone, will need at least 10 years and possibly as many as 20 to reduce its average nonperforming loan ratio to a more manageable 5%. The cost of the necessary write-offs could be equivalent to US$518 billion or almost half of China's estimated gross domestic product of US$1.1 trillion for 2001.*

Put differently, China's banking problem in 2002, when Alex Zhang came to my office, was almost certainly worse than South Korea's had been in 1998, in the trough of the Asian Financial Crisis. The difference then was that the Korean government was running out of money and had to be rescued by the IMF and the World Bank. The Chinese government, by contrast, was financially quite strong and thus had the means to bail out the banks if necessary. This was why Chinese depositors had never had real concerns about the safety of their money. Although China had no deposit insurance system such as those common in the United States, South Korea, and most developed nations, the government was always there, standing behind the banks, should a bailout be required.

We sat there, Zhang and I, talking and drinking tea. I couldn't deny it: The opportunity to buy control of a bank in China, home to the fastest economic growth in the world, was certainly appealing. Tantalizing, even. If we at Newbridge could manage somehow to take control of a Chinese bank and turn it around—as we had done with KFB—we would stand to seize a rare and, maybe, profitable opportunity. My partners and I at Newbridge had always believed that investing in banks, if properly thought through and executed, was tantamount to placing a bet on the economic future of a country. And we weren't the only ones who believed, in April 2002, that a bet on growth in China was as smart a wager as one could make anywhere on the horizon of global finance. So we thought we had to take the suggestion seriously, even if we were highly skeptical that any Chinese bank could be saved without the kind of government assistance we had received in Korea.

If. If. If! The *ifs* danced in my mind.

Chapter 2

A License
to Print Money

Au Ngai was the first colleague I picked to look at the Shenzhen Development Bank opportunity. Still in his mid-30s and carrying the title of vice president, Au had already accumulated much experience. He had joined Newbridge three years earlier than I did, after having worked at the consulting company A.T. Kearney and Bankers Trust. He had also earned an MBA from Canada's McGill University. He spoke Mandarin as well as Cantonese, the local dialect of Hong Kong and of Guangdong province, where Shenzhen is located.

A week later, Au and I went to Dorsey & Whitney's office on the 30th floor of One Pacific Place on Queensway, in the Central district of Hong Kong. Another skyscraper. Another chilly room on another warm and muggy morning. There, Alex Zhang introduced us to Zhou Lin, president of Shenzhen Development Bank.

Zhou was 51 years old. He had graduated from Nanjing University of Aeronautics and Astronautics and then obtained a master's degree from Tsinghua University, China's top engineering school. For about a decade, he had worked as a researcher in a think tank affiliated with China's State Council, the highest body of the government headed by the prime minister. Then he was appointed to be the deputy head of the Shenzhen branch

11

of the National Development and Reform Commission, a powerful body under the State Council for national economic planning, reform policies, and approvals of major development projects. Soon, he became a banker, as the Shenzhen branch manager of Guangdong Development Bank for a few years and then as the president of SDB, a position he had occupied for five years. Medium height with a somewhat stocky build, Zhou struck me from the start as warm and friendly, full of smiles. Over time we would also come to see him as being very shrewd.

We exchanged pleasantries, and then he explained that the Shenzhen city government had decided, as part of its reform policies, to bring in a qualified foreign investor to take control of Shenzhen Development Bank. And then he picked up where Zhang had left off.

"You understand banking and you understand China," Zhou said. "Your firm will be a most qualified foreign investor for SDB."

I nodded, thanked him, and Zhou then proceeded to give us a brief overview of his bank. He confirmed that the government wanted to sell about 20% of SDB's shares, which were owned by several government-controlled entities. He was candid—unusually so for a Chinese banker in those days—about his bank's bad loan problems and inadequacy of capital. After a while he threw out a number: 4 billion yuan, or about $500 million, the amount of fresh funds he believed would be needed to properly recapitalize the bank. It didn't surprise me that this bank was starved of capital. Which ones in China weren't at the time? But the magnitude still surprised me, considering the modest size of this bank.

A big number. That was my initial take. Five hundred million dollars! Newbridge was not expected to invest this amount of capital, although, at the right price, we would have been happy to consider it. No. The opportunity presented to us was to buy secondary shares from existing shareholders. Our money would not go into the bank. But Zhou pressed on, either undaunted by the amount or wishing to appear so. He had a plan.

"We've engaged Haitong Securities to do a rights issue for us," he said. "And that should raise enough capital for the bank in the stock market."

Haitong, based in Shanghai, was one of the largest securities firms in the country. A "rights issue" is a special offer to existing shareholders to buy new shares issued by the company, typically at a discount to its current market price. Only existing shareholders have the right to buy, hence the term "rights issue." The discount was supposed to make such rights issues irresistible, but to ensure success, the issuing company usually would engage a securities firm like Haitong to underwrite the offering. The underwriter had to guarantee that all the new shares would be subscribed—taken and paid for. Otherwise, the underwriter would have to buy all of the unsubscribed shares for its own book.

Zhou told us that SDB stock was trading at about five times the reported book value or net asset value (NAV). A standard measurement of a bank's financial strength, NAV quite simply is an institution's total assets minus its total liabilities. I emphasize "reported" because in light of its high ratio of bad loans, the actual NAV of the bank was likely much lower or even negative. In more developed markets, banks typically traded at no more than two times NAV. SDB's stock was traded at such lofty levels both because the Chinese stock market was highly speculative and because the Chinese economy was expected to continue its spectacular rate of growth.

China's economic growth rate had been at an average rate of 11.5% per year in U.S. dollar terms in the 10 years between 1992 and 2002. Over the next eight years, between 2002 and 2010, the average GDP growth rate, in terms of U.S. dollars (including the effect of currency appreciation), would be 19.5% per year! Of course, nobody knew at the time that China's growth would accelerate from a level that had already been improbably high.

Zhou then said the bank was planning a 3-for-10 rights issue, meaning that 3 new shares would be issued for every 10 shares held by shareholders. Most existing shareholders were expected to subscribe to the new issue. In any case, Haitong Securities, the underwriter, would commit to purchasing whatever shares that went unsold and thus guarantee the success of the planned capital raise.

The government-owned entities that would sell SDB shares to us did not plan to subscribe to the rights issue, although they would have been permitted to. If the rights issue occurred after we had bought SDB shares to become the controlling shareholder, we would not have bought the rights issue, either. On the one hand, we felt that the current stock price for tradable shares was too expensive and substantially overvalued. We could not have justified paying such a price. On the other hand, the nontradable shares that we were proposing to buy typically traded at about one time the reported NAV, which would be much more within reason, from our point of view, than the five times that tradable shares were fetching in the stock market. Consequently, the secondary shares available for us to purchase would be diluted as a percentage of total shares, but only slightly so.

It was all quite interesting, to say the least. Zhou had clearly thought things through. The hot stock market was going to rescue his bank, without any help from the government and without much dilution to existing shareholders. I had never thought this possible. Certainly it wouldn't have been possible anywhere else in the world, not with the magnitude of the capital raise he was talking about: roughly half a billion dollars to restore the health of a broken bank.

I pressed Zhou, asking what precisely gave him faith that his plan would work. He said that it had been done successfully before at SDB. A prior rights issue had also been underwritten by Haitong, which had ultimately become a shareholder of the bank by purchasing shares it had been unable to sell in the market. Apparently Haitong was confident enough to do it again.

Zhou also told us that the government controlled SDB, even though it held only about 20% of the shares. This was because the bank was a public company, listed on the Shenzhen Stock Exchange, and its stock was widely dispersed among public shareholders. The public float, meaning shares that were held by the general public, represented about 72% of the total outstanding. The Shenzhen government was far and away SDB's largest shareholder, and as such it appointed the bank's management as well as most of its board of directors. The relevant point for us was simple: If we bought those shares from the government, we would be able to control the bank, he said, just as the government had done.

This was good news, and critical to us because we had little faith that any contracts or other agreements would constitute a reliable means of guaranteeing control. China had only recently published new banking regulations, requiring that no single foreign investor own more than 20% of a Chinese bank, and no foreign investors collectively own more than 25%. Given such limits, it ordinarily would not have been possible for any foreign investor to acquire control of any Chinese bank. But SDB seemed a rare case, a unique one, because its shareholding was so dispersed, and therefore the 20% or so shares held by the government entities constituted a controlling block.

For us, this shareholding structure would prove the great revelation of the day. It was the possibility of being able to acquire control that made the deal a potential game-changer—notwithstanding all those *ifs* that served as brakes on my initial enthusiasm. A controlling position would allow us to install the right management team, adopt a necessary risk-management system, build a credit culture—to lend on the basis of the good credit quality of the borrower, and implement a strategy to revitalize and grow the bank.

It was a tantalizing thing to consider.

Still, we had to wonder. How broken was the bank? Would regulators really allow foreign investors to control a national bank? After an hour in that conference room, Au and I felt encouraged by Zhou's assurances—not convinced, perhaps, but encouraged. His plan assumed the stock market could hold its ground, and he was confident he would be able to help secure all the necessary approvals, because the Shenzhen government was behind the sale as part of its reform initiatives. On the whole, what Zhou had laid out made sense.

I also asked about other potential investors who might compete with us. Zhou told us he had contacted a number of foreign banks, including the likes of Citibank and HSBC, but none had shown interest. Presumably they had been put off by the risks of problem loans and bad asset quality. These were some of those same *ifs* that had raised doubts in my mind. Any one of these concerns might have been enough to snuff out any real interest.

We understood the concerns. By regulation, almost anywhere else in the world, a bank would have to account for the capital shortage of its banking subsidiaries whose accounts were consolidated with that of the parent. Generally, if a banking institution were to acquire a problem bank, such as SDB, the acquiring bank would have to set aside significant capital provisions—an amount from its own capital base to cover the shortage of the acquired entity. That price would often be prohibitively high.

Since Newbridge, as a nonbank buyer, was not usually required by regulators to provide capital provisions at the parent level, buying control of SDB was a more attractive proposition for us than for any banking institution. Hopefully Zhou's plan to replenish the bank's capital would work, so that Newbridge would not buy a bank with a gaping hole in its balance sheet.

Au and I were somewhat excited about the possibility of controlling a national bank in China, but we worried that the bank was too broken to be saved. We were also skeptical whether there could be a workable deal, one that would work for us as well as for the selling side. At this point, there were still too many unknowns. Nonetheless, I reaffirmed our interest.

Zhou was clearly pleased. "Now I've found someone who knows a diamond in the rough," he said, smiling, as we parted ways. He assured us that his bank would open its books to facilitate our due diligence—thorough investigation—and he promised to help us negotiate with the government for the purchase.

Zhou had struck me as smart, affable, and smooth-talking. He exuded the aura of a private dealmaker, not at all like the typically formal and reserved president of a government-controlled bank. We shook hands, agreed to meet again soon, and as he bade farewell, I felt that whatever came of this opportunity, we could do business with this man.

★ ★ ★

Since my first meeting with Alex Zhang, we'd done a bit more research about what we were getting ourselves into. SDB was a nation-wide bank, with some 200 branches in 16 major cities located in the more affluent coastal regions of China. As of late 2001, the bank employed 5,000 people and held assets worth about $14 billion. SDB was one of 15 national banks in the country. It had been formed in 1987 by combining a number of credit cooperatives and issuing stock to the public. On April 7, 1988, SDB had become the first company to be publicly listed on the Shenzhen Stock Exchange as well as the first publicly listed Chinese bank. Its stock serial number was 000001.

Banking in China was strictly controlled and regulated, more so than in most other countries. For starters, in the late 1990s, it was exceedingly difficult to obtain a banking license. In addition, interest rates were tightly regulated by the PBOC, the central bank.

Banks make money by earning a spread, the differential between lending and deposit rates. In most countries, banks are free to set their interest rates for lending and for deposits. If a bank set its lending rates too high, good customers would borrow from other banks offering lower rates. If it set its deposit rates too low, customers would deposit their money with other banks offering higher rates. So banks have to set their interest rates to be competitive. This meant the spread is usually thin.

In China at that time, interest rates were controlled by the PBOC, which set a floor to lending rates and a ceiling to deposit rates. Thus the spread was practically guaranteed by the central bank and much higher than in other markets, regardless of competition between banks. As such, banking in China was like a license to print money, provided, of course, the bank did not throw money away by making bad loans.

This "license to print money" was what strongly attracted us to the SDB opportunity—the promise of essentially controlling a money machine. It was so rare that until this moment, it had been inconceivable to a foreign investor.

Unfortunately, the state of banking in China at that time was such that banks *were* throwing money away, all too often making loans to borrowers who could not pay them back. The entire system was weak. It had weathered the Asian Financial Crisis largely because China's capital market was quite closed to the outside world, so there was practically no foreign holding of publicly traded Chinese stocks or bonds. Therefore capital flight was not an issue for China as it was for other Asian countries.

Although most Chinese banks were technically insolvent, they were kept afloat by liquidity: The inflow of deposits was typically greater than the money needed to make loans, even if individual banks could not collect some of their loans. This was almost like a giant Ponzi scheme, except that the government, whose finances had been strong, stood behind major banks so there was almost no risk they would fail. There were rare exceptions. For example, the small Hainan Development Bank was allowed to close in 1998. But even in that case, the government bailed out all the retail depositors. This was a unique phenomenon for global banking, in a country that had among the highest savings rates in the world—roughly 50% of gross domestic product or total economic output.

You didn't need a degree in finance to see that this system was untenable in the long run. Sooner or later the deposit growth rate would taper off or reverse, with the maturity of the economy and aging of its population. When that happened, the bad loan problem would catch up with the banks, making it impossible for them to make new loans and, more worryingly, impossible for them to repay their deposits. By the turn of the twenty-first century, the Chinese government was keenly aware of the weakness of its banking system and was thinking about ways to reform it.

And all of a sudden, there we were at Newbridge Capital, right in the thick of things. If we bought SDB, we would find ourselves at the forefront of China's nascent banking reforms. But our objective was simple: The control of an institution with a license to print money could be a lucrative opportunity.

In the days that followed those first meetings, my colleagues and I kept focusing on two major issues. The first was simple: No foreign investor had ever bought control of a Chinese bank. It was such a radical idea that we expected government approvals would be difficult to obtain. There was no blueprint or beaten path to follow, not even for the various layers of the government bureaucracy and regulators. The second issue involved a lack of information. We had no idea how broken SDB was, and whether it could be saved. One thing we did know: If it could *not* be saved, then we ran the real risk of losing our capital.

But then we would tell ourselves, as the worries coursed through our minds: There was no harm in checking it out. It could be, as Zhou had suggested, a diamond in the rough.

Chapter 3

The Craft
of Private Equity

In 2002, private equity—acquiring or investing in private companies by financial investors—was still a rather nascent industry, although its roots could be traced back to the corporate takeovers of the late nineteenth and early twentieth centuries. That era, often dubbed the Second Industrial Revolution, was marked by technological progress and the rise of the modern financial system, in which banks and stock markets channeled massive amounts of private capital into such megaprojects as steel making, motor vehicle manufacturing, and railway construction.

One early deal that could be considered private equity, although it was not called as such at the time, was the acquisition of Carnegie Steel in 1900 by the American financier John Pierpont Morgan. Morgan's power was so immense that he was credited with creating the first billion-dollar company in the world—United States Steel Corporation. He consolidated a vast swath of the country's railroad system under the Northern Securities Company, which had to be broken up by President Theodore Roosevelt on antitrust grounds. He was also credited with rescuing the country from financial ruin in 1907.

J.P. Morgan bought Carnegie Steel for $480 million, equivalent to about $15.6 billion in 2021 dollars. He was a man of immense financial power and capabilities, but when he died in 1913, Morgan's personal wealth was estimated to be merely $80 million (about $2.2 billion in 2021 dollars)—a respectable fortune, but pale in comparison with that of Andrew Carnegie, who received $225 million (roughly $7.3 billion in 2021 dollars) for selling his steel company. It was said that when John D. Rockefeller heard of this, he quipped: "He owned all of us and he wasn't even that rich." In view of his limited personal wealth, where did Morgan get the money to pull off such huge acquisitions?

Simply put, Morgan used other people's money, including equity capital and debt. How he did this is, essentially, what private equity is all about: pooling capital from institutional investors and making investments in, or acquisitions of, largely private companies for the purpose of generating good returns for these investors. The manager of the money, or the private equity firm, takes a cut, generally a percentage of whatever gains it can produce from its investment.

The relationship between the manager and the investor is typically structured as a partnership. The manager is known as the general partner (GP), and investors are referred to as limited partners (LPs). These investors are typically large institutions such as sovereign wealth funds, government and corporate pension systems, insurance companies, university endowments, and the like. The general partners can raise billions or tens of billions of dollars to do large transactions.

Private equity as an industry really burst onto the scene in 1988, when the partners of KKR acquired RJR Nabisco in a $25 billion transaction. The story is told in *Barbarians at the Gate* by *Wall Street Journal* reporters Bryan Burrough and John Helyar. Today, in the United States, there are more companies that are controlled by private equity firms than there are publicly listed companies.

I joined Newbridge Capital in 1998, 10 years after the RJR Nabisco deal. Before that, I had been an investment banker at J.P. Morgan in its Hong Kong office, covering China. It was only by chance that I got into investment banking and, also by chance, into private equity.

I had grown up in China and spent many years in my youth in China's Gobi Desert as a hard laborer. I had lived in squalor and had to endure starvation and other types of hardship. Those were the days when China was extremely poor. Like many of my peers who had lived through the Cultural Revolution of the 1960s and 1970s, I had never received a secondary education: During that period of tremendous social and political turmoil, all the schools were closed and remained shut for as long as 10 years. But I was lucky enough to get a chance to study in the United States when China opened up in 1980. I was able to earn three graduate degrees, including a PhD from the University of California at Berkeley. Then I became a business professor at the Wharton School of the University of Pennsylvania after a brief stint at the World Bank—my first encounter with finance.

After six years at Wharton, I became somewhat bored with life in the ivory tower and attracted to the opportunities in China, whose economy was just about to take off. In 1993, I was approached by a senior banker at J.P. Morgan, which was looking for someone who knew China. He made me an offer that has made me wonder, ever since, why bankers are paid so much more than professors. In any case, I took the opportunity, leaving the United States with my family to relocate to Hong Kong, then in its last years as a British colony.

In 1997, Hong Kong's sovereignty was handed back to China and it became a special administrative region of the People's Republic. Under the framework of "one country, two systems," it maintained its laissez-faire economy and a legal system in the British common law tradition. Today Hong Kong boasts one of the highest per-capita incomes in the world. It is an international city, with more than 10% of its population (roughly 7.5 million) comprised of foreign professionals or expatriates. Hong Kong has two official languages (Chinese and English) and is served by nonstop flights to all major cities around the globe. It has one of the world's lowest corporate and personal income tax rates (the top bracket is just 17.5%), and no capital gains or dividend tax. That, coupled with an open capital market, free flow of information, and a fully convertible currency whose exchange rate is pegged to the U.S. dollar, makes it one of the greatest financial hubs in the world, and certainly the top one in Asia.

In 1997, Asia was hit by an unprecedented financial crisis. On October 23, the Hong Kong stock index dropped a whopping 16% in one day. By the year's end, South Korea's stock market had lost 49% of its value and its currency, the won, plummeted by about 66% against the U.S. dollar. Asian countries, from Korea to Japan to Indonesia, were devastated by the financial tsunami and massive flight of capital. Despite its shaky banking system, China fared much better than other Asian countries because its capital controls had worked like a high dam keeping at bay the tsunami raging off its shores.

Just a few months before the collapse of Hong Kong's stock market in October, I had been approached by a headhunter on behalf of Newbridge. I was already restless, as I was losing interest in investment banking. The investment banking job, as I saw it, involved 80% social engineering—relationship building with clients—and just 20% financial engineering, which was the creative deal-making and financial structuring I was most interested in. I was open to approaches by headhunters, although I hadn't decided what to do. One thing led to another, and Newbridge eventually made me an offer as a partner of the firm. I sat on the offer for weeks, because I knew practically nothing about private equity, and Newbridge was a rather obscure name at the time. Then the full force of the Asian Financial Crisis hit, and the capital market went into a deep freeze, starving the investment banking business. It was then that I decided to take Newbridge's offer, thinking that in this kind of market conditions, it would be much better to be on the buy side as an investor than on the sell side as an investment banker trying to entice investors to buy.

David Bonderman, chairman of TPG and co-chairman of Newbridge, liked to say that many of us in the private equity business had stumbled into it. He himself was a Harvard-trained lawyer who had also studied Islamic law in Egypt, where he learned to speak Arabic fluently. He had taught as a law professor and worked for both the government and a private law firm before an opportunity presented itself that drew him into the private equity business. I am sure that he hadn't planned on a career in private equity, and neither had I. But when the opportunity beckoned, it looked irresistible. Bonderman was well-known in the industry, having done a number of successful high-profile deals including the acquisition of Continental Airlines and American Savings Bank, both of which were turnarounds that were eventually highly profitable for him and his investors.

Dick Blum was the other co-chairman of Newbridge. Seven years Bonderman's senior, Blum had been in the investment business throughout his career. He made partner at Sutro & Co., an investment brokerage firm, at the tender age of 30, and founded Blum Capital when he was 40.

His successful deals included CBRE, the leading commercial property management and brokerage firm in the world, and Bank of America when it was in trouble.

My co-managing partner was Dan Carroll. When we had worked on the Korea First Bank deal, our political advisors gave each of us a codename. His was Handsome Guy. Tall and good-looking, with a bachelor's degree from Harvard and an MBA from Stanford, Carroll had cut his teeth in private equity and in Asia with Hambrecht & Quist Asia Pacific, where he had been posted to Thailand, before he joined TPG and Newbridge. (By the way, my codename, which was supposed to protect my identity, was Thin Guy, which was more likely to give me away than my real name.)

Blum, Bonderman, and Carroll all lived in America but visited Asia frequently. Carroll, in particular, traveled to Asia almost every month. He and I also talked by phone frequently if a live deal was going on.

China was a difficult market, fraught with risks, particularly in the 1990s. Some early investors had gotten burned, and some had lost their shirts. Newbridge had learned its own painful lessons. Of the investments that Newbridge had made in China before I joined in 1998, Dick Blum would say: "We invested in three deals; they took from four of them!"— meaning that our joint venture partners had cheated us. The real story was much more nuanced than that, of course. But while we never lost money, we had never made much either, and that had as much to do with ourselves as our partners.

For example, we had invested in a confectionary business that had a strong national brand. It was a profitable business at the time of our investment, but it began to lose money afterward, to the bafflement of the Newbridge team. I looked into the situation after I joined the firm and discovered that we only owned a minority stake in the manufacturing part of the business, and no shares in the marketing arm. As a whole, the company was still profitable. But the profit was going to the marketing arm, which was 100% owned by our joint venture partner. Our partner could have been nicer to us by apportioning some profit to the manufacturing arm, but it was also arguable that the manufacturing part was a cost center, whereas marketing was the profit center. The fact that our own team had not structured a deal to align the interests of the parties by simply owning both parts of the business gave rise to such a conflict of interests.

Some had argued that China was different and therefore things had to be done differently. We did not agree. Yes, China was different from the U.S. market, but that did not mean one could not be successful there by following the basic rules of good investment. This experience prompted David Bonderman to say: "You shouldn't do a deal in China that you wouldn't do in Kansas." We had to follow the same standard and apply the same discipline.

But, "we live to learn," as Bonderman would say. By 2002, we felt that we had become knowledgeable enough about the market, including its pitfalls, and honed our skills to tackle it. One important rule that we had established was that we would always want to have control of the business we invested in, unless our partners could guarantee us good returns. Taking a ride with someone as a passive, minority shareholder was just too risky.

All of our partners were intrigued by the possibility of controlling a national bank in China. But they all were skeptical about the quality of the bank, and about whether we would be able to put together a deal that would work for us—allowing us to turn the bank around and make money for our investors.

Newbridge had put its name on the map in Asia by acquiring control of Korea First Bank from the Korean government in 2000, a deal that was hailed by the *Wall Street Journal* as a "bold stroke" and a "bellwether" of Korea's banking reforms. I had worked with Dan Carroll, Paul Chen, and Daniel Poon as a team for the KFB deal, supported by other partners and professionals from time to time. For the SDB transaction, the makeup of the core team was different, consisting mainly of Au Ngai, Daniel Poon, and Ricky Lau, in addition to myself. Ricky Lau is a native of Hong Kong but, like Au, he speaks both Cantonese and Mandarin Chinese. Daniel Poon is also a native of Hong Kong; his Mandarin was yet to be fluent, although he had no problem understanding it. All of us were familiar with China and its market.

The Island Shangri-La Hotel is a five-star property located in the center of Pacific Place, a sprawling complex of hotels, office buildings, and shopping malls, adjacent to Hong Kong Park, an 860,000 square foot (80,000 square meter) spread of trees, ponds, meandering paths lined by all kinds of vegetation, and a zoo. The park is surrounded by high-rises of Hong Kong's Central district. The hotel is a popular place for businesspeople and tourists alike. It offers a variety of restaurants—Western, Chinese, and Japanese.

It was there in Shangri-La, an appropriate name for the secret project we had started, that we met again with Zhou Lin on April 19, 2002, for breakfast in a private room. Alex Zhang had booked the room so we could avoid prying eyes. Whatever was to come of these discussions, even a hint that such a deal was in the works would have drawn interest from journalists and competitors. My Newbridge colleagues Au and Poon joined the meeting. We affirmed our interest in SDB, and then we talked about money—the price we would have to pay for the block of shares to be sold by the government. Both parties understood that the public market price for SDB stock would not be the benchmark for our purchase; we would be buying "legal person" shares, which were nontradable in the stock market.

The stock market in China was still young. The Shanghai Stock Exchange had opened less than a decade before, in December 1990, with a good deal of fanfare. The Shenzhen Stock Exchange, where SDB was listed, had its groundbreaking the following year, although SDB had been traded on a so-called over-the-counter market from the time of its minting as stock number 000001. According to the rules of the time, when a company went public, only the shares offered to public investors could be freely traded. These were referred to as "tradable shares." Shares owned by the pre-IPO owners, which typically were state entities, were referred to as "legal person" or "nontradable" shares. Legal person shares could be bought and sold, but only on an off-market basis, through private transactions, rather than in the open market.

Because Chinese stock markets were far from mature and rules were still lacking or were poorly enforced, stock prices remained highly speculative, and some were blatantly manipulated. More often than not, the market price of a stock bore no relationship to the fundamentals of the issuing company.

As discussed in my first meeting with Zhou, SDB's stock was trading at about five times its reported net asset value (NAV, also called *book value*). The actual NAV was likely to be much lower or even negative, depending on how bad SDB's loan book was. Elsewhere in the world, bank stocks were usually traded at about two times book value, some more and some less. Five times reported book value was almost unheard of outside of China, which gave you some idea how speculative China's stock market was at the time. If, as we suspected, the true NAV of SDB was lower, the stock price would represent a multiple even higher than that.

I said to Zhou that banks in foreign countries were typically priced at a multiple of their tangible NAV, excluding such intangibles as goodwill—an accounting measure of the difference between the price paid for an asset and its tangible net asset value. If the bank's reported NAV did not fully take into account the provisions and write-offs required to cover bad loans, then it would need to be adjusted. To us there seemed one reasonable approach: We would calculate the purchase price as a multiple of the *adjusted* tangible NAV.

Zhou said he understood. He acknowledged that the bank's NPLs were indeed under-provisioned, but said he worried that if we adjusted NAV for under-provisioning, the adjusted NAV would be negative, which would render any multiple of the adjusted NAV meaningless— the multiple of a negative number is infinite but still negative. A negative price would mean the current shareholders would have to pay Newbridge to take their shares. And that, of course, would be absurd. On the other hand, legal person shares were typically traded at more or less equal to *reported* NAV. He proposed that we either pay a price on the basis of the *reported* NAV, or a high multiple of adjusted NAV, assuming it would still be positive.

We were generally okay with his suggestions, but it was impossible to talk about pricing without really knowing SDB's assets and what the true NAV was. Zhou was anxious to work out a deal, and he could see we were knowledgeable about all these issues. We agreed on the spot, before we had paid the breakfast bill, that we would sign a letter of intent giving Newbridge the exclusive rights to a deal, but we would defer the details, including any discussion of pricing, to a later date. Among other things, we needed to conduct due diligence on the bank, to check out the quality of its assets and other aspects of the bank's operation.

On April 26, less than a month after our first meeting, Zhou and I executed a confidentiality agreement and a letter of intent. As the president of SDB, he represented the bank, and I represented Newbridge

Capital. The letter was both an expression of intent and a roadmap for both sides. It set forth a few key points:

1. Newbridge would purchase shares of SDB from the government (or, more specifically, from a group of government-controlled entities) for a consideration to be determined. "In principle," the agreement read, "Newbridge is prepared to pay between the investment cost by the Sellers and the current NAV for the shares." The idea was that the price had to cover the seller's cost of having bought those shares in the first place, but no more than the stated book value.

2. In order to recapitalize the bank, as Zhou planned, SDB would do a rights issue of "no more than 3 new shares for every 10 shares held by the existing shareholders."

3. Newbridge would replace the sellers' existing representatives on SDB's board and would have the right to nominate its own members as well, such that directors appointed or nominated by Newbridge would represent the majority of the board.

4. Finally, this would be an exclusive arrangement. "The Sellers and the management will undertake not to engage in any similar discussions with regard to the sale or purchase of the shares held by the Sellers to any other party."

All our potential deals at Newbridge required code names. For the KFB deal we had chosen Project Safe. Now, in the early days of our negotiations for SDB, we called this one Project Sterling.

To take the next step, the team would require the approval of Newbridge's investment review committee, or IRC. Made up of all the partners at the firm, including our two co-chairmen, the IRC oversaw all our important investment decisions. Everyone on the committee was intimately familiar with banking because of our past experiences. They knew that to obtain control of a national bank in China would be a highly usual opportunity, but nobody expected it to be easy, nor did anyone harbor any illusions about the quality of the bank's assets or whether the deal would actually be doable. But we would never know without checking it out. The internal memo that the team wrote to the IRC laid out, among other things, the difficulties of obtaining the necessary regulatory blessings.

"Regulatory approval from the highest level of the Chinese government is required," we wrote, given the unprecedented nature of such a transaction in China. The memo continued:

> *The process will be long and difficult as the deal would have to be ultimately approved by Zhu Rongji, the prime minister himself, and all major departments of the Chinese government, including the central bank, securities regulators and Ministry of Finance. Therefore the probability of an eventual deal remains rather low.*

We would need to keep our expectations low as well. The committee unanimously agreed that the team should continue to explore.

Chapter 4

A Special Time and Place

As I sat in a Hong Kong skyscraper and imagined negotiating the landmark purchase of a control stake in a Chinese bank, I thought of how many roads had been traveled to bring the right people together for this moment. And there was the road China had traveled from the late 1970s to the century's turn—it had journeyed from dire poverty to relative prosperity and even, in certain corners, to wealth. And my own path was intertwined with China's.

In 1975, when I got out of the Gobi Desert, China's gross domestic product stood at $163 billion—roughly on par with Canada, whose population was only 2.5% of China's. Bicycles ruled the roads of the nation's metropolises. The service sector hardly existed; after all, you were supposed to serve yourself in a proletarian society. It was difficult to obtain anything beyond the basics. Even by the time I arrived in Hong Kong in 1993, China was still a poor and developing country. Its GDP had tripled to $440 billion, but that was just one-16th that of the United States, and a tenth of Japan's. Per-capita income was $377 in China, a tiny fraction of that in the United States ($26,000) and Japan ($38,000). But things were rapidly changing.

By 2002, China's GDP had reached $1.47 trillion, nine times what it had been in 1975. Cities had mushroomed in size, and cars began to fill the streets. Anywhere you looked, you saw a forest of construction cranes in every direction, and a network of brand-new highways, bridges, and airports connected the major cities.

What had happened? You didn't need to be a political scientist, or a banker, to grasp the basics.

First, Deng Xiaoping, China's paramount leader, had made the calculation to open China to other parts of the world, and to private enterprise. It had begun slowly in 1978, but Deng had lit a spark, and had done it in a very clever way. "Let some people become rich first," he said. That ignited the fervor of entrepreneurship and freed the animal spirits latent in the Chinese population. Game on, as they say in the West.

At five-foot-two, Deng may have been short in stature, but he was as tough as nails. He suffered numerous setbacks in his long revolutionary career through war and peacetime, but he had managed to bounce back each time. Born in 1904, he went to France at age 15 to work and study. Then, he studied in Moscow before returning to China to work for the Chinese Communist Party (CCP), eventually becoming a leader in the Red Army, which was fighting with Chiang Kai-shek's government. During the last civil war (1946–1949), he played a key role in the Battle of Huaihai, in which the CCP forces, 600,000 strong, annihilated 800,000 of Chiang Kai-shek's troops, sealing Chiang's fate on the mainland (Chiang fled to Taiwan) and paving the way for the CCP's takeover of national power. Deng was only 45 at the founding of the People's Republic of China, but he was already a core member of its leadership.

In 1966, Mao Zedong launched the Cultural Revolution, and Deng was purged from the party and the government for being a "capitalist roader." He advocated for economic policies that were more pragmatic and tolerant than Mao's. Driven by ideology, the Cultural Revolution and its violence rapidly spread nationwide. Mao's loyalists attacked those deemed insufficiently supportive of the movement, including members of Deng's family. His eldest son was paralyzed after jumping out of a window to escape from the torments of the Red Guards—anti-establishment student organizations. Deng himself was banished to do manual labor on a tractor factory in Jiangxi province, a remote area 900 miles south of Beijing. Then, in 1973, Mao brought him back to the power center and appointed him executive vice premier. At the time, Zhou Enlai, the prime minister, was gravely ill, so Deng oversaw the country's economic affairs. It didn't take long for his policies to once again clash with Mao's, and Mao once again removed him from power for being, in Mao's words, "never willing to repent to the day of his death." By then, 1976, Mao was dying.

Mao must have found Deng's independent mind equally brilliant and maddening. Even after removing him from the position of executive vice premier, Mao still thought of Deng as "a rare talent difficult to find," as tough as "a steel mill," and "a needle wrapped in cotton." But Mao also complained about him, saying, "He is said to be hard of hearing; yet, in every meeting, he sits as far away from me as possible." In economic policies, Deng was always a pragmatist. Whereas Mao was ever vigilant against any signs of capitalism, Deng took a different point of view when it came to bolstering China's economy. He was well known for saying, "It doesn't matter if a cat is black or white, as long as it catches mice." All the remarks, either by Mao or Deng, were well publicized in official newspapers at the time, so all in the nation were made aware of them.

Deng returned to power in 1978, two years after Mao's death, and became the paramount leader, although he only took the title of vice premier. He immediately embarked upon market-oriented reforms and opened China's doors to international trade, investments, and exchanges of students and scholars. Mao's shadow hung over Chinese politics, so Deng's policies were bold. He wasn't afraid of taking risks—"If the sky falls, the taller guys would hold it up," he liked to say, in self-deprecating reference to his own short stature.

Deng brought one game-changing policy after another, pushing China in the direction of a market economy and integration with the global economy. He normalized China's diplomatic relationship with the United States in 1979 and visited the country in the same year. In a meeting with Deng, President Jimmy Carter requested that Beijing be flexible on emigration. As Zbigniew Brzezinski, Carter's national security advisor, recounted in his memoirs, Deng leaned forward and asked, "Are you prepared to accept 10 million?" The American president quickly dropped the subject.

Deng died in 1997. He was hailed as the chief architect of China's economic reforms and his policies have lived on.

As the country opened up, its foreign trade soared. China's exports to the United States alone almost doubled within five years, from $51.5 billion in 1996 to $102 billion in 2001, the year China became a member of the World Trade Organization (WTO). Admission to the

WTO provided a further impetus for market reforms domestically, and it led to greater international trade. China became a manufacturing juggernaut, especially in coastal regions, which quickly became factories for the world. All of these developments propelled the economy, which sustained a double-digit rate of growth.

But this momentum disguised a major vulnerability in the economy. The nation's banking system, which had fueled the rapid economic growth, had also accumulated a massive number of bad loans, so much so that most of the banks were technically insolvent. It would be difficult for the country to sustain its growth without cleaning up its banks. As part of its membership agreement with the WTO, the country was also obligated to gradually open up its banking sector to foreign competition, against which domestic banks, weak and outmoded as they were, stood little chance. A major banking reform was called for, and it looked more pressing by the day.

This was the context in April 2002, when I first heard about the possibilities for Shenzhen Development Bank. Today it may all seem quaint, a small sliver of history, but in 2002, it felt like a miracle—a new China, the kind of place where a wild idea like the one Zhang and Zhou were pushing seemed worth looking into.

★ ★ ★

The idea of selling control of a national bank to foreign investors was conceptualized as a critical piece of the broader reforms. It wasn't clear to us if the specific plan for SDB had been the brainchild of the Shenzhen government, but it did embrace it. Our team drove to Shenzhen on the second Friday in May with Alex Zhang, for our first real engagement with some local officials.

I already knew the city. I had first visited it in 1981, before it was much of a city at all. Then, there were no skyscrapers, and it was hard to imagine the construction boom that would surge through Shenzhen. By the time I arrived in 2002, a modern skyline was emerging, dotted with construction cranes, although it had yet to rival Hong Kong in any real sense. Shenzhen was gleaming, a forest of tall buildings. The population had skyrocketed to 7.5 million from merely 68,000 in 1981, a 110-fold jump in a span of just 20 years. Nowhere on Earth at any time in history had a major city sprung up as quickly. Everything seemed new. In some ways Hong Kong looked almost shabby by comparison.

How did this come about? Shenzhen was the brainchild of Deng. In 1980 he decided to undertake a bold experiment, designating Shenzhen one of four "Special Economic Zones," which were allowed to adopt market-friendly policies that were much more liberal than anywhere else in the country. The idea was that if the experiment failed, it would be confined and would not affect the rest of the country. If it succeeded, it would set an example for the rest of the country. And so it was that the old, sleepy fishing village with a vast expanse of paddy fields became a Petri dish for market liberalism, China-style. Just like that, it was soon flooded by tidal waves of economic activity that changed the landscape beyond recognition in just a few years.

We drove on highways and tree-lined boulevards of multiple lanes. Shops lined both sides of the street, and the sidewalks were crowded with shoppers and pedestrians. Young women were dressed in skirts in bright, tropical colors. Traffic was heavy.

Zhou Lin met us and brought us to a restaurant in a hotel to meet with Song Hai, a vice mayor, and Liu Xueqiang, the city's deputy secretary general. Both of them greeted us warmly. Then we sat down for dinner.

Song was about 50, with a slightly receding hairline. He was an alumnus of the Beijing Institute of Foreign Trade, the same college I had attended. He had graduated a year ahead of me, with a degree in Spanish—likely not a language he had used much after graduation. He had also earned a PhD in economics from Nankai University. As vice mayor, Song's responsibilities included the financial affairs of the city. Local banks would fall under his purview.

Liu Xueqiang, the city's deputy secretary general, was a few years younger than Song. He came across as more of a scholarly type. I later learned that he was a published author of short stories and essays.

Song emphasized the Shenzhen government's general interest in foreign investment. He spoke of the laissez-faire policies and the measures they had taken to bring in foreign investment. When it came to SDB, he wanted us to know that the city government recognized the benefit of foreign ownership. Reform was high on the agenda for the nation and Shenzhen had led the way. It was not entirely surprising that the city government would take this bold step. Nobody at the table talked about the weakness of the bank. They were clearly aware that it was troubled, but I didn't know to what extent they understood the challenges it faced. In any case, they must have concluded that bringing in a reputable foreign investor would be good for the bank and for the city.

Both Song and Liu told us that the city government had long resolved that privatization was the way to go to reform state-owned enterprises. Otherwise, they said, the troubles with these companies would be the government's responsibilities and would consume the government's limited resources, including tax revenue. If these businesses were privately owned, the government would still be the beneficiary of the profits they made by way of collecting taxes, but would have no responsibility if they lost money or failed.

The message made sense, and it was refreshing. Not every local government in the country thought the same way. Shenzhen was certainly at the forefront of market-oriented reforms. What surprised us was the way they hammered the point home. Together Song and Liu shared a rhyme that they said summarized the policy and attitude of the Shenzhen government toward foreign investors:

> You invest, we welcome with open arms;
> You make money, we collect taxes;
> You commit crimes, we make arrests;
> You go bankrupt, it is none of our business.

We all had a good laugh. The saying was reflective of the government's thinking in selling the controlling interest in SDB. At least it was a fitting phrase for the city of Shenzhen.

I thought it also served as an apt way to describe much about the direction that China was taking, although Shenzhen, as a Special Economic Zone, was probably way ahead of other major cities. Under the old system before economic reforms began in 1978, the government had owned everything and had to provide loss-making firms with financial assistance and subsidies. In the new market economy that Shenzhen was promoting, the government realized that firms were likely to be better managed in private hands than under government ownership.

"We seek your presence, but we don't seek ownership," the Shenzhen officials had told us. Or, as one line in Song and Liu's little ditty went: "You make money, we collect taxes." It was clear that the city government considered SDB a liability and wanted it sold to investors who could turn it around. In China's first Special Economic Zone, it made perfect sense. In the broader Chinese context, it was of course quite bold, as bold as the policies that had given rise to the new Shenzhen.

Song told us the government would do everything possible to facilitate the transaction. There was one key condition: He wanted Newbridge to commit to keeping the domicile of the bank in Shenzhen. They were concerned we might move the SDB headquarters to Shanghai, where most large Chinese banks were based.

Shanghai was attractive for its history, its status as China's financial center, and its more central location, but Shenzhen offered its own advantages. Whereas China's corporate income tax rate was typically 33%, Shenzhen-based companies enjoyed a preferential rate of a flat 15%. Ironically, SDB had benefited only minimally from the low rate, because its money-making branches were in higher-tax regions such as Shanghai and Guangzhou; its Shenzhen-area branches were in the red. Since taxes were paid locally, SDB could not offset losses from its profits. Taken together, SDB was paying an effective tax rate in excess of 40%.

I felt a good chemistry with Song and Liu. From our brief time together, these officials had struck me as capable and open-minded. They were genuinely supportive of bringing in an experienced foreign investor for SDB.

★ ★ ★

We needed to lock down the deal. Zhou Lin was also eager to push the process forward. The next step was to sign a formal sales and purchase agreement for the shares of SDB. My colleagues and I met with Zhou in the office of Dorsey & Whitney on June 11. The purpose of the meeting was to negotiate and reach agreement on the price we would pay for the SDB shares.

We had to be cautious. When we acquired Korea First Bank a couple of years back, we had spent more than a year negotiating with the Korean government, chiefly over how we, as investors, would be protected from bad legacy loans. Eventually, the government agreed to provide guarantees for the existing loan book, thereby shielding investors from potential losses. In the case of SDB, we knew that the government would not take responsibility for existing bad loans because even though the government had appointed the management, it had only owned less than 20% of the shares. There was no reason for the government to bail out the bank, giving more than 80% of the shareholders a free ride. Besides, there were regulations governing the disposal of state-owned assets. They could not be sold below the bank's net asset value (NAV), regardless of whether the reported NAV was inflated. We had to buy the bank as it was, taking the risks of potential write-offs from the nonperforming loans (NPLs).

But the circumstances were also different. In 1998, when we began to negotiate the acquisition of KFB, South Korea was in the throes of a deep economic crisis with no respite in sight. Nobody knew how long the down cycle would last and how deep it would get. We would not have been able to acquire the Korean bank in a deep recession without the financial assistance from the government.

But China was a completely different ballgame. Its economy was booming. Its economic growth rate accelerated to 9.1% in real terms in 2002, up from 8.3% in 2001. In a growing economy, a properly managed bank can grow out of its problems by tightly controlling the risks of new loans. In addition, the bank might be able to tap into a hot stock market to recapitalize itself without significant dilution to existing shareholders. These were the reasons we were willing to consider buying a troubled bank in China without any financial support from the government.

We agreed quickly that we would pay a multiple of the adjusted NAV. Both sides were fully aware that the book value of the bank was not clean, and that it needed to be adjusted to take into account all those under-provisioned NPLs. The two sides were also aware that the adjusted book value might be negative. We would have no way of knowing for sure until our due diligence was complete.

Commercial banks make money by capturing the difference between the average interest rate at which it lends money and the average interest rate at which it collects deposits (or borrows money in other ways). The difference between the two average interest rates is called the spread. The bank wants to make sure that the spread is positive. But a positive spread doesn't necessarily mean that the bank makes money. There are two major costs to consider. One is operating cost: compensation for its staff, paying rents and utility bills, and other expenses. The other is credit cost, which refers to loan losses. If these costs are greater than the spread it earns, the bank loses money.

Regardless of whether a bank makes or loses money, it is still obligated to cover its borrowings—including paying back its deposits, which are effectively loans made by depositors to the bank. Depositors fully expect to get their money back under any circumstances. If a bank loses money, say, by not being able to collect on loans it has made, it generally has to dip into its own equity capital in order to make sure its deposits are paid back. That equity capital is the same as the NAV, the difference between the bank's total assets and all of its borrowings, including deposits. The amount of NAV relative to a bank's total assets is a good indicator of its resilience or strength. But it isn't unusual for a bank to under-provision (for potential loan losses) for its loan book. If so, the reported NAV would not reflect the true equity value.

However, how much to provision against a questionable loan is often subjective. If it is beyond any doubt that a borrower won't be able to pay back the loan, such as in the event of bankruptcy, then writing off the loan shouldn't be controversial. If, however, a borrower's financial conditions have deteriorated but it isn't yet bankrupt, then it is a matter of judgment how much provision should be made against the loan. Should it be 20%, 30%, or 50%? It is for this reason that the NAV of a bank can be an elusive number—difficult to assess and potentially manipulated. Often a third party must be brought in to make an independent determination. A buyer of a bank will typically want to make its own determination as to whether there is such under-provisioning, and if so, the buyer will want to adjust its reported NAV.

We had known that this next phase in our talks would be difficult, and we went back and forth for a while. Initially Zhou proposed that we pay four times the adjusted NAV. That was out of the question. Even the best banks in the world traded at no more than two times book value. I counterproposed 1.5 to 1.7 times adjusted book value, citing market precedents and other comparables. We spent the whole morning and all of lunch trying to narrow our differences. Eventually, both sides compromised; we agreed tentatively to a price representing 2.5 times adjusted book value—at which point Zhou said he had to check with Vice Mayor Song for his guidance.

He strolled into a different room to call Song. For a few moments, waiting there with my colleagues, I felt good about things. There was the adrenaline rush that came with progress, and the knowledge that we were headed for something special. Then, less than 15 minutes later, Zhou returned to the conference room with a sobering piece of news: Song was insisting we pay five times adjusted book value.

Really? I thought. Five times? This was 25% higher than Zhou's initial unrealistic salvo. It was a nonstarter, and I told him so. But beyond the wildly inflated figure, the whole episode also cast doubt on Zhou's ability to negotiate. Why had we spent the morning hashing out terms, only to have them thrown out by the vice mayor? I could understand that top local officials might feel the need for a high number, to share with their bosses. But we could never justify such a price. Zhou must have known as much, and just to be sure, I told him it was unacceptable.

"Don't worry," Zhou said with the hint of a smile. "The adjusted NAV will be either zero or negative anyway." Five times zero was still zero; the multiple, he was saying, would ultimately make no difference.

Anyone who has done a major deal in China over the past few decades—or even attempted such a thing—will recognize the feelings: dizzying optimism, tempered quickly and often by questions and potential obstacles that crop up along the way. Or by strange moments such as these. We had known such wild swings of fortune in Korea as well. But in China, the peaks and valleys were extremes, and the nightmare scenarios could make you dizzy. Listening to Zhou, I kept thinking that he ought to have known how broken his bank was, and how many bad loans were on its books. And if the real book value of SDB was indeed zero or negative, then any multiple would be meaningless and irrelevant anyhow. In that event, why were we even having this conversation? We would have to find some other ways to price the shares.

I told Zhou we would think about it overnight, but I had made up my mind already. If the adjusted book value was in fact negative, the whole discussion would change. Obviously the government wasn't going to give us the bank, or pay us to take it. Equally clear, calculating a price by multiplying a negative number would make no sense. In such a circumstance, we would return to the reported NAV as the basis for pricing, before the downward adjustment.

Zhou believed we had to have a headline number of five times in the agreement between us, as a figure that would be attractive to the officials involved. I did not want to get stuck on this issue but in some sense it *was* the issue. Not just an obstacle, or a bump in the road. Zhou told us he would have to speak with regulators and government agencies to help garner support for the transaction. And so he left, somehow positive and encouraging about the whole thing. He seemed to be on our side and he understood our concerns. He told us he was confident about winning the necessary approvals, although he also stressed that those would require a good deal of work. And then he was off for a flight to Beijing.

We understood that Zhou was well connected, that he knew many people in various government agencies. He represented the bank in question and he had been given a mandate by the Shenzhen government to win approvals from Beijing. More important, despite the gap in expectations between the vice mayor and our team, we trusted Zhou. He knew which buttons to push, and we all hoped he could gather the support needed to make sure that once we filed for approvals, they would all be given. This much we knew: Without his prewiring the system, the deal was unlikely to get very far.

Chapter 5

Dancing with the Wolves

The architecture of Shenzhen's Wu Zhou Guest House was modern and grandiose. Even the name suggested immensity: *Wu Zhou* is Chinese for *five continents*, and *guest house* is a misnomer leading you to imagine a quaint and quiet setting. But *five continents* fit, as the name for this ambitious, sprawling structure. From a distance, Wu Zhou looked like the upper body of a giant robot, with a pair of wings spread wide. The body of the robot was a 10-story building, and the ground floor an enormous lobby that led to several banquet halls, including a ballroom that could host 2,000 guests. Branching from the center were two wings, which held several hundred guestrooms each. Its reddish frame was studded with large glass windows, and the entire building was surrounded by gardens, flowerbeds, and trees. The hotel complex had been built by the city government in 1997, and it remained in government hands. Today, almost every international five-star hotel brand has a property in Shenzhen, including Four Seasons, Ritz Carlton, St. Regis, and Shangri-La. But when we arrived, in 2002, Wu Zhou was one of only a few high-end places to stay.

We came to know the sprawling complex well in those first few months we were navigating the SBD deal. I rarely spent the night—Shenzhen was only an hour's drive from either my home or office in

Hong Kong—but we visited often. Under the framework of "one country, two systems," a visitor from Hong Kong to Shenzhen still had to navigate border controls on both the Hong Kong and mainland sides, just as one would when crossing an international border. Vehicles issued license plates by both Hong Kong and the mainland were allowed to cross after clearing the border and custom controls on either side.

On June 21, 2002, SDB sent a chauffeured car to bring my colleagues and me from Hong Kong to the Wu Zhou Guest House, where, in one of those giant conference rooms, SDB and the Shenzhen officials had planned a ceremony for the signing of a *Framework Agreement for Sale and Purchase of the Shares of Shenzhen Development Bank.*

The agreement was fairly simple: Newbridge would acquire all the shares of SDB held by the city government. Those shares represented about 20% of the total and constituted a controlling block, due to the fact that the rest of the shares were widely dispersed. Newbridge would have effective control by having the right to appoint a majority of the board of directors and new management. The price was set at five times *adjusted* book value but the multiple, 5, was shown in brackets, indicating that it remained subject to negotiation.

It was not a final agreement, but it would give Newbridge the exclusive right to conclude a deal to acquire the controlling interest in SDB under the mutually agreed framework. The agreement legally obligated the sellers to sell to Newbridge Capital, but we would retain the right to step away if we uncovered any unpleasant surprises in our due diligence work.

Our lawyers had felt making any clause in the framework agreement legally binding was not necessary at this stage. We had yet to conduct our due diligence—thoroughly investigating the operations and financials of the bank—to our satisfaction. I insisted that the document be generally binding so that under no circumstances could the seller walk away from the deal, or *a* deal, since some details, such as the price, were yet to be agreed.

We had learned a lesson about nonbinding agreements during the negotiations for Korea First Bank. We had signed a memorandum of understanding (MOU) with the Korean government on December 31, 1998, which gave us a four-month period of exclusivity to negotiate and conclude the transaction. The nonbinding nature of that MOU had brought a lot of headaches and anxiety. We had known at every moment of that four-month stretch that the seller could walk away from the transaction, even after months of hard work and millions of dollars in transaction costs. In the end, things had worked out, but I was determined not to let that experience—and all the angst that had come with it—repeat itself. We would have enough stress as it was. This framework agreement with Shenzhen was made binding on the sell side, although our obligation to close the deal remained subject to due diligence to our satisfaction. In other words, we could walk away, but they could not.

On the morning of June 21, we were met in the guest house by Zhou Lin and Liu Xueqing, the deputy secretary general of Shenzhen. I had e-mailed the final version of the framework agreement to Shenzhen before we got on the road. For some reason, however, the version placed on the table in the Wu Zhou conference room was the wrong one—I always check the wording of a contract before I sign it. I did not have time to wait for the correct version to be produced, as I had to return to Hong Kong for a flight to San Francisco. We decided that Zhou and I would sign the signature pages for the camera, with Liu and my colleague Au Ngai as witnesses. Zhou would sign on behalf of the selling shareholders, and I would for Newbridge Capital. I left after the signing, and Au stayed behind to sign the correct version.

And so it was that two months and two weeks after that first meeting, we had concluded the framework agreement for a deal to buy a Chinese bank. We were exhilarated. Eyes wide open to the rough road ahead, but excited. As that odd bit of saying went, *You invest, we welcome ...*

★ ★ ★

The signing at the Wu Zhou Guest House was a big step forward. The framework agreement set us on a path to something we couldn't have imagined only a few months before. But it was still just one step. Many details needed to be worked out, particularly the price. The whole idea of Shenzhen selling the controlling interest in SDB, as the rest of the deal, would still be subject to approval by multiple levels of authorities, including in Shenzhen, Guangdong Province (of which Shenzhen is one of the cities), and Beijing. Approval from the city government seemed a foregone conclusion, but we weren't so sure about the provincial level. And as for Beijing, the deal would have to be reviewed by multiple regulators, including the central bank, the securities regulator, and the Ministry of Finance. Ultimately, it would need the assent of the State Council—the central government—and the prime minister himself.

For all these reasons it was hard to get too excited, but we were cautiously optimistic.

The Shenzhen side was making every effort to keep the deal a secret, but word was beginning to get out. A somewhat obscure Chinese newspaper, *21st Century Talent News*, had become the first outlet to break the news that SDB might be looking for a foreign partner. "Mr. Zhou Lin recently confirmed the rumor that SDB is talking with foreign banks to bring in a foreign investor," the paper reported. "SDB may become the first publicly listed joint stock commercial bank with foreign ownership." The reporter was skeptical about whether a foreign investor would be a good thing for the bank, or for China's banking system. He concluded his article with something more like a fable than a news story, with an ominous tone. It appeared under the heading "Dancing with the Wolves":

After many turns and twists, China finally joined the WTO. Many industries including banking will gradually open up to foreign competition. So some people cry wolf, some other people say dance with wolves, and yet some other people say they want to turn themselves into wolves. Regardless, all have the mentality of being a sheep.

The reporter went on to recount a fable about a wolf hoping to lure a sheep out of its pen. The wolf told the sheep that there was an oasis nearby, and that it was very beautiful but only had enough grass for two to share. The wolf suggests a plot: If the sheep will help the wolf kill the other sheep, the wolf will eat grass and they can share the oasis. The sheep agrees and leads its flockmates one by one out of the ringed fence. Then, when all of them have been eaten by the wolf, the sheep walks out of the pen and waits for the wolf to lead the way to the oasis. The reporter for *21st Century Talent News* continues the story:

But the wolf stares at the sheep with his bloodshot eyes, bares his teeth and says, you think there is such a good thing in the world? How can a wolf change his true nature of eating sheep but to eat grass? Looking at the smiling and confident expression of Zhou Lin, this reporter suddenly remembered this story.

I laughed this off, but I recognized that this strange fable reflected the mentality of the time for many in China: Foreign investors like us were probably up to no good. We were the proverbial big, bad wolf. But as the author noted, Zhou saw things differently. And so did I. The truth of the matter, I believed—borrowing another metaphor from the animal kingdom—was that most Chinese banks were like lazy pigs, content to sit in their pens and feed off customers without providing adequate services. They could get away with it because there was no competition. The pigs were doing just fine, and they wouldn't change their ways unless prodded. Chinese banks might not need a wolf, I thought, but a border collie would certainly do them some good.

This wolf-and-sheep conversation spoke to those seismic changes in China: Deng's economic opening, the new economic zone in Shenzhen, and China's accession into the WTO in 2001, when the country had begun to allow limited foreign entry into its banking market. By mid-2002, the framework of the WTO had yet to make a dent in China's banking market. Foreign bank branches were only allowed to engage in corporate banking, not retail banking. Furthermore, they were restricted to opening one branch per year, and even then only if approved by the central bank. Therefore, in the foreseeable future, foreign banks could pose no meaningful competitive threat to Chinese banks.

But under the WTO agreement, China was committed to gradually opening up its financial services sector to foreign competition. The Chinese leadership saw this as a critical part of China's economic reform and particularly the reform of its financial system. What Newbridge proposed to do—to take over control of a national bank—was not part of China's obligation under the WTO agreement. If it happened, it would be China's own radical initiative in reforming its banking sector.

What were the odds of Newbridge Capital winning all those regulatory approvals? China's new banking regulations specifically required that a foreign investor in a Chinese bank be a "financial institution" with total assets of at least $20 billion, three years of profit history, and investment-grade credit ratings from one of the major international ratings agencies, such as Standard & Poor's or Moody's. Newbridge was a private equity firm that could hardly meet all these requirements. On the other hand, the whole concept of foreign control of a national bank was new. There had to be some policy breakthroughs for our proposed acquisition to happen.

After signing the framework agreement, I flew as planned to San Francisco on the night of June 21. I was with my daughter LeeAnn, shopping in the Stonestown Mall, when my phone rang. It was Au. Au was not someone who overtly showed his emotions. He calmly told me that Zhu Rongji, the prime minister, upon returning from a business trip, had reviewed our proposal to acquire control of SDB and delegated the matter to Wen Jiabao, vice premier, and Ma Kai, secretary general of the State Council.

Even though Zhu Rongji did not indicate his personal views one way or the other, we thought the fact he did not object to the proposed deal was significant enough. To allow a foreign investor to control a national bank was a huge deal. We had not expected him to approve the idea immediately, and he could have smothered it out of hand. His delegating the agreement review to the senior members of his cabinet to study and make their recommendations was tantamount to giving the green light to proceed further, although eventual approval remained an uncertainty. Nonetheless, I was pleasantly surprised that things had moved so quickly.

By 2002, Zhu Rongji had been the prime minister for four years, but he had been the top government leader of China's economic and financial affairs since 1993 when he became the first-ranking vice premier (there were several other vice premiers). He had a reputation for being a strong reformer. He closed down inefficient state-owned enterprises and allowed private firms to flourish in competitive industries. He personally negotiated China's accession into the World Trade Organization in 2001. At this time, he had shifted his attention to banking reforms, especially the large and dominant government-controlled banks, although how he would tackle this giant problem was not yet clear to market observers. Under his oversight, some medium-sized banks were already transformed to joint-stock companies whose shareholders included private investors, although most remained controlled by the government.

One thing was certain: Zhu Rongji's reform agenda was far from over.

★ ★ ★

In his personality and leadership style, Zhu Rongji was a bit of an oddball in the Chinese government. He was admired and feared by his ministers because of his bluntness, attention to detail, and impatience with bureaucratic incompetence. He graduated from Tsinghua University, the best engineering school in China, I and then worked as a junior official in the National Planning Commission in his early career. In 1958, his outspokenness got him into trouble when he criticized Mao's Great Leap Forward—a radical collectivization of agriculture and reckless expansion of industries that ended up creating famine and hardship. Zhu was expelled from the CCP and, during the Cultural Revolution, he endured a five-year exile in the countryside doing manual labor. But after Deng Xiaoping returned to power, many who had shared Zhu's plight were rehabilitated. Zhu was readmitted into the Party. He quickly rose through the ranks because of his competence, capabilities, and his no-nonsense style (or perhaps in spite of it). He became nationally known after becoming the mayor of Shanghai, where he earned the nickname One-Chop Zhu for simplifying the approval process for businesses. In China, a *chop* is a stamp used to sign things. Zhu reduced many bureaucratic layers of red tape to one chop, the official seal of approval.

I first met Zhu, along with many other people, in 1990, when I was a professor at the Wharton School. An alumnus, a successful businessman from Singapore, had funded a Wharton executive education program in Shanghai. The program was run in cooperation with Shanghai Jiaotong University, one of the best institutions of higher learning in China. At the time, Zhu was Shanghai's mayor. The Wharton School appointed me as the director of the program. The two-week program was taught by Wharton professors flown in from Philadelphia and attended by the rising stars of the city government—promising mid-level officials and executives of state-owned companies. Some of the graduates of the program eventually rose to prominence, including becoming members of the Politburo, the top decision-making body in China. These students were eager for Western ideas in business and management. Some of them spoke English; others used simultaneous translation provided for the program.

Wharton held a dinner party to mark the conclusion of the program for all the faculty members and the participants. Zhu also attended. During the dinner, we requested his support for another program of the same kind to be held the following summer. Without hesitation, he held up the pen we had given him as a token of appreciation, and loudly said, for all to hear, "I hereby give my approval for a repeat of the Wharton program next year!"

I resigned from the faculty of the Wharton School in 1993 and became a J.P. Morgan banker and chief representative in China. By then, Zhu had become first vice premier. Soon after, the prime minister was rumored to have suffered a heart attack, so the responsibility of state affairs rested largely on Zhu's shoulders. He would go on to become the prime minister in 1998.

In December 1995, I went with Sandy Warner, J.P. Morgan's chairman and CEO, and some other colleagues to visit with Zhu in the central government compound, Zhongnanhai. It was a time when top Chinese leaders were quite accessible to visiting business leaders, and our meeting was arranged by the Ministry of Finance, which hosted the J.P. Morgan delegation. As the visitors wanted to learn about China's policies directly from the top, the Chinese leaders were also keen to welcome foreign businesses and learn from the visitors.

Adjacent to Beijing's Forbidden City, the ancient seat of imperial power, Zhongnanhai was once part of the imperial gardens. Our meeting took place in a lakeside building named Zi-guang-ge, or Purple Light Pavilion. *Pavilion* was an understatement, as the building was palatial. It was first built around 1500 in the Ming dynasty and rebuilt in 1700 in the Qing dynasty. In late Qing dynasty, around the mid-1800s, the emperor received foreign dignitaries there. It remained a venue for Chinese leaders to meet with foreign visitors.

"I did my homework before your visit," Zhu said, after formally welcoming Warner, "and I learned that J.P. Morgan isn't Morgan Stanley. I wouldn't have met you if you represented Morgan Stanley."

In Chinese, Morgan Stanley was known as "Big Morgan," and J.P. Morgan was "Morgan Bank," making it easy to mix up the two. In fact, both firms were born out of the House of Morgan, but they were split in 1933 by the Glass–Steagall Act, a Depression-era law that separated investment banks from retail and commercial banks. J.P. Morgan had only returned to investment banking in the 1990s, when banking companies began taking advantage of loopholes in the Glass–Steagall Act, which was repealed in 1999.

"Mr. Jiang called me earlier and asked me if you were Morgan Stanley," Zhu continued, referring to our next scheduled meeting with Jiang Zemin, China's president. "He wouldn't see you either if you were."

The origin of Chinese leaders' ire against Morgan Stanley was a pair of economic predictions made by Barton Biggs, then its chief strategist.

In 1993, Biggs had declared himself "maximum bullish" about the Chinese economy. In that year, the very first batch of Chinese companies went public on Hong Kong's stock exchange, opening the gate for many more to come in subsequent years. Hong Kong's benchmark Hang Seng Index had already more than doubled from about 2,500 in 1990 to about 6,000 in 1993. Following Biggs' "maximum bullish" call, the Hang Seng almost doubled again, reaching more than 11,000 in 1994. This bullish market had helped propel the initial public offerings of Chinese companies into the stratosphere as international investors chased them. Then in 1994, in another research note, Biggs warned that the Hong Kong stock exchange was now a bubble that "was about to burst." The market dropped precipitously and by early 1995, the Hang Seng Index had dropped to below 7,000. Morgan Stanley, Zhu complained to us, had manipulated the market and "made a handful" of money in the process. I did not think that anyone on J.P. Morgan's team believed that Morgan Stanley had deliberately manipulated the market for its own gains. Nor was there any point for J.P. Morgan's chairman to rise to Morgan Stanley's defense against China's vice premier. In fact, the J.P. Morgan team was probably gleeful that our major competitor had drawn the wrath of the Chinese leaders. It was in fact all a big misunderstanding because shortly afterward, the Hong Kong stock market resumed its ascent and reached a peak of about 16,000 by July 1997. But Zhu's remarks reflected his style of being brutally blunt.

Confident and self-assured as he was, he could also be persuaded to change his mind.

In 1997, a few years after my first meeting with Zhu, my team at J.P. Morgan helped advise Baosteel, the largest steel mill in China, on obtaining credit ratings from Moody's and Standard & Poor's. The exercise was to prepare Baosteel to raise capital through what was called a Yankee bond, a bond denominated in U.S. dollars. It was a very successful endeavor: Both agencies gave Baosteel a credit rating on par with China's sovereign rating, or the same as its ministry of finance. (Moody's later revised the rating one notch down because its internal policy changed, to not give any firm a rating the same as the sovereign's.) A good credit rating allows the issuer to pay a lower interest rate for its borrowings. But at the time, for a Chinese company to issue such a bond in overseas markets required Zhu's approval.

Soon, the response to Baosteel's request was handed down from Zhu's office with his handwritten remarks. Zhu had approved the Yankee bond, but he said that, in conjunction, Baosteel should also acquire some other steel companies that were insolvent.

At the time, Zhu was leading the reform of China's bloated state-owned enterprises, and his priority was to solve the problems of these inefficient and loss-making firms. Baosteel was not one of them. In fact, the reason that Baosteel was able to obtain such high credit ratings was because it was in excellent financial condition with a strong balance sheet. If it had acquired or merged with weaker firms, Baosteel's own credit rating would suffer and it would not be able to make the planned bond issuance.

The vice premier's stand put Baosteel's management in a bind. The chair of Baosteel, Xie Qihua, was the only female CEO of a large steel mill in the world. She was known, quite fittingly, as an iron lady—a capable and decisive leader held in high regard by other executives. She impressed the foreign rating agencies with what she had achieved as the CEO and her knowledge of the company. But in the Chinese hierarchy, a company's CEO was of course many layers removed from the vice premier and she did not feel like contradicting her ultimate boss by directly petitioning him to reconsider his decision. She asked me if I, as an outside advisor, would write a letter to Zhu to explain that a successful issuance of the contemplated bond depended on the credit quality of the issuer, which would have been damaged if it had acquired a loss-making firm.

I agreed to do it but had no expectation that the vice premier would change his mind. I knew top leaders anywhere usually did not want to admit that they had made a mistake. To my surprise, a few weeks later, Baosteel received fresh approvals from the vice premier's office to issue the bond without conditions. I was impressed not only with Zhu's efficiency and decisiveness, but also his willingness to accept reason and different viewpoints from a humble and obscure banker. Zhu had the reputation of being a no-nonsense reformer for good reason.

The top leader of the country in 2002 was President Jiang Zemin. Jiang and Zhu had worked together since the late 1980s, when Jiang was the CCP chief of Shanghai and Zhu was the mayor. In 1989, when Deng Xiaoping tapped Jiang to become the head of state, Zhu succeeded Jiang as Shanghai's party chief, until he too took up his current position in the central government. Now, as prime minister, Zhu had the support of Jiang to push forward with his reform agenda.

I had got a glimpse of Jiang's style during that trip to Beijing with Sandy Warner in December 1995. Whereas Zhu came across as sharp and rather stern-faced most of the time, Jiang was all smiles and greeted us warmly. But he too was quite straightforward. After we were seated, Jiang opened the conversation by pointing at his chair and saying, "Do you see the chair I am sitting in?"

We looked at his chair and at each other, not knowing why he drew our attention to his chair. For a moment, I thought it was an antique or something of great value—there are many pieces of antique furniture in the Great Hall of the People, where the meeting was held—but it did not look like anything special.

Then he said, "Anyone sitting in this chair would not allow Taiwan to become independent."

Ah, that was what he was getting at—he was referring to his own position as the head of the state. I was sure his remark surprised and puzzled our chairman and his entourage of American bankers, as many of them likely had no idea what he was referring to. It was a time later referred to by historians as the Third Taiwan Strait Crisis.

Since Richard Nixon's first visit to China in 1972, the U.S. government had followed a "one-China" policy—meaning that Taiwan is recognized as part of China, although under a different government. The mainland and Taiwan were separated in the last year of China's civil war (1946–1949) when Chiang Kai-shek's troops fled to the island. Beijing has vowed to reunify with Taiwan eventually. When China and the United States normalized diplomatic relations in 1979, the United States severed its official relationship with Taipei and closed its embassy there. Consistent with the one-China policy, Washington had not granted visas to any of Taiwan's leaders to visit the United States. This understanding was broken in 1995 when the Clinton administration granted a visa for Taiwan's president, Lee Tenghui, to visit his alma mater, Cornell University. Beijing had already been alarmed by some remarks that Lee had made suggesting his leaning toward Taiwan's independence. Beijing regarded his visit as Washington tacitly condoning his policies. That set off great tensions, as China responded by conducting military exercises, including firing missiles (albeit without warheads, as was later revealed) into the

Taiwan Strait between July and November of that year and in March of the following year.

So what Jiang meant when he met with us was that any Chinese leader in his position would not allow Taiwan's independence. He just chose the occasion of the visit by a group of clueless American bankers to deliver the message. Of course, he knew our chairman was visiting for business purposes, but I think he probably broached the subject because he thought we were concerned about the tension across the Taiwan Strait and felt it necessary to explain China's position.

To drive home his point, Jiang proceeded to tell us a story. Like Zhu, he had also received an engineering degree, but from the elite Shanghai Jiaotong University in 1947. (It was no accident almost all top Chinese leaders had engineering degrees, as the field had attracted the best and brightest.) His first job offer was to be an electrical engineer at Taiwan's Sun Moon Lake Hydropower Station. At the time, he was already an underground member of the CCP, and the People's Liberation Army was fighting with Chiang Kai-shek's Nationalist forces for the control of China. Jiang's party bosses instructed him to stay in Shanghai in anticipation of a victory over the Nationalist regime.

Taiwan, Jiang wanted to make clear, was part of China. "I could have been an engineer living in Taiwan today," if it were not for the decision of the party to keep him in Shanghai, he said.

As I was listening to him, my thoughts momentarily drifted to a meeting I had in 1989, with Sun Yun-suan, the former prime minister of the Republic of China, as the government in Taiwan was called. (The leaders of Taiwan also claimed sovereignty over all of China.) I was visiting Taiwan as a Wharton professor and went together with a few other scholars to meet with Sun as he was well-known as one of the architects of Taiwan's economic takeoff in the 1970s and 1980s. His speech was thickened by the familiar accent of Shandong Province, where my parents had come from, and Sun told us that his first job after the Nationalist government moved to Taiwan had been as the head of the Sun Moon Lake Hydropower Station.

Sitting in that reception room in the Great Hall of the People with Jiang Zemin, I thought that if history had taken a different turn, his life could have overlapped with that of Sun Yun-suan in that hydropower station in the middle of Taiwan. How unpredictable life could be.

But Jiang's main interest was not Taiwan on that day. He quickly changed the topic to electronics, a subject dear to his heart as he had been the minister of electronic industries between 1983 and 1985. He talked about cutting-edge technology. The subject matter was so technical that his translator struggled. Breaking with protocol, Jiang shifted from Chinese to English. He talked passionately about new breakthroughs in electronics, but his conversation was laced with so many technical terms that I wasn't sure if his audience was able to follow his train of thought.

The impression I got from that day's meetings was that he and his government were serious about driving market-oriented reforms and industrialization. And they considered foreign investment and participation in the Chinese economy as critical to their reform and growth agenda.

Seven years had passed since those meetings when Newbridge started negotiations to acquire control of SDB. Now, in 2002, the talk in China was about bringing foreign banks and investors into China's banking industry in order to modernize the industry and follow "international practices" in making changes to the economy. It seemed that Newbridge had arrived at the right place at the right time.

Chapter 6

Rolling Faster

It seemed that the ball had started rolling fast. The fact that Zhu Rongji had not raised any objections to our proposed deal for SDB surely augured well for the other necessary approvals. But we also knew that this was only the first step. We had now, I felt, been allowed to compete in the race, but there would still be many obstacles to overcome to reach the finish line. There were innumerable other agencies and regulators both at the local and the central government levels we would need to work with before it was over.

From the other side of the Pacific, wandering in the Stonestown shops with my daughter, my mind drifted back to Hong Kong. I kept thinking of how sharp and well-connected Zhou Lin had to be to have gotten the proposal so quickly to the prime minister's office. Zhou was deftly navigating the Chinese bureaucracy, no question. Others might have started from the bottom, submitting the framework agreement to local governments and regulators in Beijing, waiting at every turn for the relevant approvals. But that route would likely have left the deal bogged down with various agencies, simply because there were no rules or precedents to follow. If the matter could be handled from the top, on the other hand, then the various officials down the line would know to move faster. This was Zhou's method: He worked on the top officials from day one.

As my daughter shopped, I was on the phone telling my partners of what I had just learned in the phone call from Au. We would all soon congregate in Aspen, Colorado, for our annual offsite gathering and we would have plenty of time to think about and plot the next steps for the deal. When LeeAnn and I walked out of the mall, we were both very happy, I with the knowledge of the good news about SDB and she with a small shopping bag of fashion accessories.

★ ★ ★

China's central bank, the PBOC, had requested detailed data and information from Newbridge, and over the next few months we worked intensively to meet its requirements. We needed, first, to share details about Newbridge's credentials. I drafted a paper laying out who we were, our background, and what we had accomplished, especially in the bank-ing sector. With the help of our team, I worked day and night to prepare the document in both English and Chinese, with Zhou Lin offering cri-tiques and edits along the way. I knew it was imperative for us to differ-entiate ourselves from other investors by demonstrating our track record in owning and turning around banks.

In the papers we produced, we described Newbridge as a "strategic financial investor" and reviewed in detail our role in taking control and resuscitating Korea First Bank. Harvard Business School had just published a case study on our KFB transaction—a nice thing for us, as third-party recognition carried more weight and value than anything we might say about ourselves. We also included a U.S. Congress–sponsored study on the acquisition of American Savings Bank, a deal that Bonderman and his team had done in the late 1980s, to further bolster our case.

Meanwhile, SDB engaged the investment bank of Salomon Smith Barney, an arm of Citigroup, as its advisor. Salomon's team, led by Francis Leung, knew us well, and for the most part its analysts accepted the data we provided. Bespectacled and soft-spoken, Leung was known for his pioneering work in bringing Chinese companies to overseas public markets during his tenure as CEO of Peregrine, the Hong Kong–based investment bank he had co-founded. Peregrine had gone under during the Asian Financial Crisis, after failing to recover some $265 million—about

50% of its capital—that it had committed to an Indonesia taxi company, Steady Safe (a truly unfortunate name, as it was neither). Leung had not been directly responsible for the fiasco, though, so the demise of Peregrine had not dented his credibility in the eyes of Chinese regulators.

The PBOC was prudent and did thorough work. Its officials wanted to know how we were regulated in the United States. They requested that we obtain a letter from the Federal Reserve to certify that we were in good financial standing. It was difficult to meet with the request; as a private equity firm, we were not regulated by the Fed. Nonetheless, our name was not unfamiliar to the relevant Fed officials; we had dealt with them in connection with the acquisition of KFB.

One of the lawyers who had handled the KFB transaction, Jack Murphy with the firm Cleary Gottlieb, spoke with several Fed officials and obtained a letter from the Fed's Board of Governors that basically said that the Fed had no regulatory oversight in this case. Dated August 8, 2002, the letter read:

> Newbridge Capital does not own or control a U.S. bank. Therefore, Newbridge Capital and its affiliates are not subject to supervision or regulation by the Federal Reserve under the Bank Holding Company Act. The proposed acquisition by Newbridge of an ownership interest in SDB would not cause Newbridge to become subject to banking supervision or regulation in the United States because SDB does not conduct banking operations in the United States. For these reasons, the Federal Reserve has no bank supervisory or regulatory role with respect to the proposed transaction.

The letter was nonetheless helpful, as the Fed acknowledged that it knew us and was raising no red flags. This, we felt, could give comfort to the PBOC. Another Cleary Gottlieb lawyer, Rich Lincer, met with a PBOC representative in New York City to provide background information about Newbridge, particularly in the realms of investing in and turning around banks.

What remained unsettled—and would for some time—was the fundamental question of price. What would we pay for the shares of SDB? While the framework agreement had referenced five times the adjusted

NAV, that number would remain meaningless until our due diligence was complete and the adjusted NAV was determined. While we waited, both the Shenzhen government and the PBOC wanted the parties to provide an estimate of the final price. At Zhou's request, I submitted a letter on August 6 to the Shenzhen government, under the heading "Commitment by Newbridge regarding the investment in SDB."

In the letter, Newbridge committed to paying a price in the range of 0.8 to 1.5 times the *reported* net asset value in the event that the computed price based on the previously agreed formula of five times the adjusted NAV was lower than this range. The commitment was intended to give comfort to the government side that even if the adjusted NAV was zero or negative, as feared, it would still receive a price in line with the prevailing market of about one time the reported NAV. The condition to the price range was for the Shenzhen government to commit to providing SDB with preferential policies with regard to debt collection and capital market operations, including the planned issuance of new stock by SDB to raise capital.

That same day, SDB submitted a document marked "top secret" to the PBOC. The letter endorsed Newbridge as a qualified investor and reported that "the government of Shenzhen and Newbridge mutually agreed to a minimum price of 0.8–1.5 times of the reported NAV of SDB." These documents would become critical at a later stage in the deal.

On August 10, SDB also submitted a report to the PBOC outlining its view that Newbridge would be the best choice for the bank. We were not privy to the report itself, but Zhou shared the highlights with me over the phone. I was impressed by its thoughtfulness and eloquence, no doubt penned by Zhou himself, who was a good writer. The report described our background and track record in bank deals and presented us as investors capable of turning the bank around. It was as good and positive a document as we might have written ourselves.

Three days later, we received the news that the PBOC had given its in-principle approval for Newbridge to acquire the controlling stake in SDB. One more hurdle cleared.

★ ★ ★

We had been very careful, from the beginning, to keep everything related to SDB strictly confidential. If the deal itself was unprecedented, so too would be the interest among reporters, competitors, and others in the financial community. You could never know from which corner opposition, competition, or trouble might arise. The sheer number of people involved in the process meant leaks were hard to prevent.

So I was not surprised when I got a call from a Bloomberg reporter in late July, while driving with my colleagues Au Ngai and Ricky Lau to the SDB headquarters. The reporter went straight to the point, no preamble whatsoever: Was Newbridge looking to invest in SDB?

I had decided to dodge such questions, if and when they came. To the Bloomberg reporter's query I said simply, "We don't comment on rumors."

The drive from Hong Kong's central business district to Shenzhen typically takes about an hour, including the time to cross the border. When we arrived at the SDB headquarters, an SDB staffer showed me an article in the *Economic Observer*, a Chinese publication, that said that HSBC, J.P. Morgan, and Newbridge were competing for a stake in SDB. This was obviously why the Bloomberg reporter had called. The rumor had already been picked up by several major Western media outlets. Dow Jones reported:

> *HSBC Holdings PLC, J.P. Morgan Chase & Co. and Newbridge Capital Inc. are rival bidders for a stake in China's Shenzhen Development Bank, the weekly* Economic Observer *reports.*
>
> *The local government-controlled bank has recently been approved to sell all of its legal person shares, the report says. Armed with the new approval, the bank is currently negotiating with three rival bidders to determine the best outcome for the planned share placement.... The scramble for a stake in Shenzhen Development Bank, one of only four banks listed on mainland stock exchanges, again pits HSBC against Newbridge Capital. In 1999, Newbridge beat HSBC for a stake in South Korea's Korea First Bank.*

The *Economic Observer* may have been onto the story, but it had some key facts wrong. For one thing, I knew from Zhou Lin that neither J.P. Morgan nor HSBC was competing with us, although they had been approached at one time or another. SDB was known to be a weak bank, if not a basket case. Major banks would balk at the possibility of investing in such a troubled institution. Nor did I think that anyone else had figured out that a foreign investor could buy control of SDB, given the limits on foreign ownership. In any case, we had already won the exclusive rights to any deal, under the framework agreement, so rumored competition did not faze us.

In a way, though, the *Observer* story was good for Newbridge. It served as a status booster. If SDB was perceived as a desired object, something the big banks were fighting for, well, that could only make us look good. We now had distinguished company. It also suggested that we were not the only ones crazy enough to take the risk (although we were). None of this really mattered to us, but such perception mattered to the parties on the sell side and to regulators.

In response to the press speculation, SDB suspended trading of its stock on July 30, in accordance with the trading rules that required it to clarify any news or rumor that might disturb its stock price. SDB issued a terse statement saying there was no disclosable information that it had not already shared with regulators or the press. Since we were still negotiating and a deal remained uncertain, the bank did not think it had an obligation to say anything further. Then the stock resumed trading the next day.

★ ★ ★

SDB's headquarters was in the Luohu District of Shenzhen, about a 15-minute drive from the border, on the south side of a major thoroughfare named Shenzhen South Road. The building was quite new, completed in 1996, and stood 34 stories high but seemed taller. It was shaped like a gigantic sail, meant to symbolize smooth sailing and good fortune. All the window glass was tinted the color of pink champagne, a strange hue that reminded me of the pages of the *Financial Times*. The FT's publishers, I thought, would have been proud to own the building.

On that day of our first visit, July 30, my colleagues and I had a productive session with Zhou Lin and other senior officers, although few knew why we were there. We went through some of the documents we planned to submit to regulators. Then, we took a tour of the marble-colonnaded business hall with a 40-foot high ceiling and the customer reception area with service counters in the large banking hall. At the deep end of the hall, a giant oil painting hung on the wall, illuminated by soft lights shining from above: A huge banyan tree, the symbol of

The champagne-colored building is the headquarters of Shenzhen Development Bank, based in Shenzhen, China.

Shenzhen, stood by a lake into which waterfalls from surrounding mountains dropped, and a group of 12 oxen were grazing under the tree. The painting, by a Hong Kong artist named Liu Yuyi, was entitled "The Eternal Lake of Fortune." How appropriate, I thought.

The colonnaded Business Hall at SDB's headquarters.

This oil painting hung on the back wall of SDB's Business Hall. The artist is Mr. Liu Yuyi from Hong Kong. He did this painting specifically for SDB and completed it in 1996, when the new SDB headquarters building opened.

In some ways, SDB's main office was like any other bank in China, or, for that matter, anywhere in the world at that time. Customers lined up in front of windows that sat above a high counter to wait for service. This was the old-style, impersonal banking designed to keep a distance from customers—to intimidate them, perhaps—as well as to deter any would-be robbers. The modern design we had adopted in Korea First Bank was to have bank officers and customers sitting across a desk in an open area as they discussed business. SDB was definitely old-school in this sense. But in other ways, it was very much an edifice built to show-case modernity. No old-school bank would have pinkish windows or a sail-shaped design.

★ ★ ★

The central bank, after having given us the preliminary green light to pursue the transaction, forwarded the documents to the China Securities Regulatory Commission (CSRC) and the Ministry of Finance (MOF) for their review, respectively. The MOF approval was needed for any sale of state-owned assets—in this case, the SDB shares owned by the Shenzhen government. The CSRC did not raise any issues at this stage. Zhou Lin had anticipated some difficulties in obtaining MOF's support, but on the evening of August 16, he called to say that MOF had also given its in-principle consent. Once again, I thought things were moving in the right direction at a fast pace. The PBOC submitted the papers from three regulators along with recommendations to the State Council for approval. The file was sent to the office of Wen Jiabao, a vice premier who raised no objection and passed it on to the prime minister's office on August 27.

Four and a half months after the idea had first surfaced, the leaders of China were reviewing our deal for the second time, now with the in-principle consent of all the relevant regulators. The approval process may have seemed circular, but that was how it worked. Since the deal would be ground-breaking, the regulators wouldn't want to examine it without knowing that the top-level officials would even entertain such an idea. Although the top-level officials, including the prime minister and the vice premier, signaled their willingness to consider it, they still needed to see the opinions of the relevant regulators to affirm all was proper. Even after the second round, the approvals remained preliminary. There

would be more steps to take and more rounds to go through. The process, though complicated, minimized risks of surprises for the regulators
as well as for ourselves.

★ ★ ★

Prime Minister Zhu Rongji had initiated China's banking reforms in
1999 by creating four asset management companies (AMCs) that would
take on the nonperforming loans (NPLs) that were being carved out of the
four largest state-owned banks.

Ordinarily, if a bank's bad loans become so large that the potential losses from them are greater than its capital base, the bank will fail.
In China at that time, a bank could be insolvent, but it would not fail,
because it was implicitly backed by the full credit of the national government. However, without being able to collect on the vast amount of
money it had loaned out, a bank's ability to make fresh loans would be
constricted. To put it simply, removing NPLs from a bank's balance sheet
would be the financial equivalent of unclogging a drain, so that everything
was able to function properly again.

The purpose of AMCs was to buy NPLs from the biggest banks, taking on the risk of managing these problematic loans and clearing the way
for the bank to get back to lending money. It was a tested strategy, one
we ourselves had employed a version of when managing the bad loans
made at KFB.

However, Zhu Rongji knew the measures would be futile if the
banks continued to make risky loans. His next step was to bring in foreign investors and then turn the government-owned banks into public
companies, listed on overseas stock exchanges. The idea was that foreign
investors would introduce international best practices including robust risk
management systems, and that the requirements of overseas stock markets
would impose market discipline upon them.

In the next decade, all of China's Big Four banks—Industrial and
Commercial Bank of China, China Construction Bank, Bank of China,
and Agricultural Bank of China—would be listed on overseas stock markets and boast a roster of foreign "strategic" investors, such as Bank of
America, Goldman Sachs, and the like. In 2002, the full scale of this reform
agenda was not yet apparent, although a few small, city-wide commercial

banks had begun to sell a small percentage of their shares to foreign financial institutions. However, control of these banks would remain in Chinese hands.

By comparison, our proposed transaction was unlike anything China had seen. SDB would remain a Chinese bank but be controlled and run by Newbridge, a U.S. investor. Would the Chinese leadership really go that far in reforming its banking sector? The approvals we had received so far already represented a major policy breakthrough in Chinese banking.

I asked Zhou Lin what he thought the odds were that we would be able to get to the finish line.

He gave a classic Chinese reply. The "weather forecast," he said, was "partly cloudy." No sun as yet, but not overcast either—meaning hopeful but by no means certain.

★ ★ ★

We were winning over a range of allies and advocates. The highest office in Shenzhen belonged to the party secretary, a post held by a lady named Huang Liman. Zhou Lin told us she was strongly supportive of the SDB transaction and would lobby Beijing for necessary approvals. On August 29, Zhou took me, my colleague Au Ngai, and Francis Leung of Salomon to have dinner with Huang in a government guesthouse. We arrived at about 5:30 p.m., unusually early for dinner, even by Shenzhen standards, where people typically ate at about 6 p.m. Vice Mayor Song and Deputy Secretary General Liu Xueqiang also joined the dinner.

Huang Liman was in her mid-fifties. Plainly dressed, she had a casual and easy demeanor. She had received an engineering degree and had worked in the ministry of electronics industry when Jiang Zemin, China's president, was the minister. She had been working in the Shenzhen government for 10 years and had been made the party chief only the previous year.

The conversation meandered from problems with SDB to the Shenzhen government's policies for developing the city. Huang spoke of the strong consensus in Shenzhen to sell the stake in SDB to a foreign investor. "This decision is strategically very significant," she said. "The government is determined to exit from firms, to be a referee but not to be a player." Such talk was popular among government officials at the time to explain why the government was privatizing state-owned companies. The idea was that in a fair competition, the government should not be both a referee, as regulators, and a player, as the owner of state-owned companies. By stepping out of businesses, the government could create a more competitive and efficient market.

Then she added, with no prompting from us, that the authorities would give Newbridge free rein for controlling SDB: "After we transfer the shares to you, we will let you manage the bank. The government will only support. We will not interfere."

She was speaking as if the deal had been done. I thanked her and assured her that Newbridge had the experience, ability, and confidence to turn around SDB, and to increase its market value and competitiveness. I stressed that we would bring in a world-class management team and international best practices in banking.

She seemed happy with what I said and volunteered that she would personally lobby Zhu Rongji's office and the prime minister himself if necessary. Our dinner went on for two hours before we bade her farewell.

On the following Monday, September 9, Zhou called me at 9:30 in the morning. Huang had made her case, and Prime Minister Zhu had approved, in principle, our proposed transaction. This was a necessary step, and not by any stretch a final one. But once again the prime minister had proved true to his reputation—hard-working, efficient, and decisive. Now the road was open for us, although we still had some distance to go. I wrote down in my diary on that day: "I was delighted ... immediately called David Bonderman, Dick Blum, Dan Carroll, and others to share the good news." All were pleased to hear it, although we all knew there remained a lot of work to do.

By protocol, the official document bearing the preliminary approval of the prime minister and the State Council had to go to the provincial government in Guangdong, and then to the city government of Shenzhen. It would take time. Meanwhile, market speculation about a potential sale of SDB shares to a foreign investor had gone from occasional rumor to nonstop chatter. The Shenzhen Stock Exchange asked SDB to issue a public clarification, as public companies are often required to do when stock price–sensitive information is being rumored.

On September 11, SDB formally announced that it was indeed engaged in talks with foreign investors about selling a stake. In a public statement released to the press, it confirmed the discussions but still did not disclose the name of the suitor.

> *Our bank has recently contacted and talked with several foreign institutions on cooperation and introducing foreign strategic investors. The move is to implement the Shenzhen municipal government's strategy to promote reforms and development of state enterprises in line with the new conditions after China joined the WTO.*

> *Relevant government authorities have approved our plans in principle to bring in foreign strategic investors. We are moving forward with our negotiations. But we cannot confirm the final results at this moment.*

The announcement naturally aroused still more curiosity. Since Newbridge had already been mentioned as one of the suitors, journalists came calling. There was, to put it mildly, a lot of buzz. I decided to speak with a few reporters I knew well from both the Western and Chinese media, on a background basis, to prepare them for the news. We wanted to make sure the message and the story were not mangled by rumor or error, and I was careful not to disclose any information that might be considered sensitive or confidential.

Meanwhile, we had our own investors to worry about. On September 18, we sent a memo, marked "Strictly Confidential," to roughly 50 of our investors, known as limited partners (whereas Newbridge itself was the general partner), to provide a brief summary of the deal we were working on:

> *As mentioned in our last quarterly report, SDB is a mid-sized China-based commercial bank with approximately $16 billion of total assets and over 190 branches spread across major cities around the country. Its current loan portfolio is comprised primarily of corporate loans, with its consumer lending business currently in development.*

> *Newbridge entered into exclusive negotiation with the major shareholder of SDB last April, which resulted in the execution of a sale and purchase agreement on June 21, 2002. This agreement anticipates that Newbridge will acquire approximately 20 percent of SDB from the Shenzhen Municipal Government and is conditional upon completion*

of due diligence, to Newbridge's satisfaction, and the receipt of all necessary regulatory approvals. Upon closing, Newbridge will appoint a majority of the Board of Directors and will control the appointment of the bank's senior management. The exact investment amount will be determined on the basis of adjusted net book value, determined through Newbridge's due diligence process. Having said that, we expect the Fund's investment to be approximately $120–150 million and that our entry price will be a fraction of where the stock is currently traded, although our stake will take the form of non-listed "legal person" shares.

We laid out more detail and closed the memo with a note about the deal's potential as a landmark transaction for China:

Taken together, the policy breakthroughs embodied in this approval are hugely significant as it marks the first time since the founding of the People's Republic of China, in 1949, that a foreign investor will have ever controlled a Chinese bank with a nationwide franchise. In fact, under the World Trade Organization framework, foreign banks are not permitted to conduct local currency retail business in China until 2007 and are not permitted to establish more than one branch per year, and then only if approved by the central bank.

We hammered home the unique aspect of banking—a license to print money due to interest rates controls by the central bank, and the enormous potential in the market:

Our investment thesis is that SDB is uniquely well positioned to tap into China's rapidly growing retail banking market, particularly in the area of mortgage and housing loans.

How much money could we expect to make from such an investment? At this point, we did not know. That required much more work in connection with our further investigation and analysis. Eventually, we would need to build a computerized model to inform us of potential outcomes as the underlying assumptions varied.

Our original agreement had stipulated that Newbridge would take control of the SDB board of directors immediately after the central government's in-principle approval. If we were to become a controlling shareholder, then we would own whatever risks the bank was taking from this moment on. We wanted to send in experienced bankers and immediately install a robust risk management system to ensure that the bank stopped making loans of dubious quality.

Indeed, the risk was substantial. SDB's NPL ratio—as a percentage of total loans—was reported to have reached 14.84% for the year ending 2001. SDB was in agreement with us on this. On September 16th, the bank submitted an "urgent report" to the Shenzhen government explaining why Newbridge should be invited to take control of the bank even before the transaction closed:

> *Even though the acquisition has already been approved [in principle] by the leaders of the State Council and relevant authorities, there may still be risks and surprises which we don't want to see before the dust settles: First, if there are different views and the competitive requests by our peers, the leaders of the State Council might change their mind and call off this deal. Second, as rumors fly and spread, the probability of risks and accidents within the bank has greatly increased. Between January and August of this year, SDB's nonperforming loans have increased by more than 2.9 billion [yuan] ...*

We were not yet shareholders of SDB, but we agreed with the bank's management that we would effectively take control in a two-step process: the Shenzhen government would appoint directors, and then those directors would transfer their proxies to Newbridge representatives. There followed a brief dispute about the legality of the proxy arrangement, pitting the legal department of the city government against the head of the Securities Regulatory Office of Shenzhen. Ultimately, it was resolved that the board of SDB would authorize Newbridge to do things, without giving us the proxy to vote on behalf of board directors.

It was a small thing, a hiccup really. And I could understand the caution exercised by these government officials—an aversion to potential legal risk. But the episode left me concerned that it might mark a shift in how the Shenzhen side would make decisions as we moved forward. Previously, we had negotiated effectively with Zhou Lin, who represented SDB and communicated with government agencies in Shenzhen and Beijing. This channel had streamlined things, and no doubt it had smoothed and hastened the process. Now it seemed many more people might be involved, and that would likely complicate matters at every turn.

To comply with this latest decision of the city government, Newbridge's lawyers drafted a letter for SDB to authorize our representatives to take control of credit decisions. The authorization was quickly approved by SDB's board, and it allowed Newbridge to form a Transition Control Committee consisting of Newbridge representatives and their nominees. The committee's responsibility was to supervise credit decisions. As a practical matter, this new body would be in no position to approve any new loans. We would not want to be taking responsibility for any new loan until we really were owners. Instead, the transition control committee would only have the authority to reject loan applications. That way, we could only err on the conservative side, by vetoing a loan approval. This construct was acceptable to SDB's board and management.

Now that the in-principle approvals had been obtained, SDB was ready to issue a public announcement. We worked together with SDB staff to draft a brief statement, which was approved by the stock exchange on September 27th:

Announcement by Shenzhen Development Bank Company Ltd.

The relevant departments of the government have given the approval for our bank to bring in a foreign strategic investor—the U.S.-based Newbridge Capital. The parties are actively making preparations for due diligence and shall make an effort to complete the transaction as soon as possible. We will make further disclosures in accordance with the progress and relevant regulations.

Those three sentences were good enough for the hungry financial press. The next day, the news was everywhere. Despite all the earlier speculation, the formal announcement still rocked the market. I received a flurry of calls from reporters, but we decided not to make any further comments. The market—and the press—were beginning to realize the significance of the transaction. "China Approves Deal for Newbridge Capital," read the headline in the September 30 edition of the *Wall Street Journal*. And below that: "U.S. firm could be first foreign institution to head a mainland bank." The article drove right to the core of what made this deal different:

> *As long as the U.S. investment firm's stake remains under 25 percent, Shenzhen Development Bank would be regarded as a local bank rather than as a foreign-controlled one, meaning Newbridge Capital could exercise effective control but avoid the restrictions that limit the business scope of foreign banks in China.*

Meanwhile, *BusinessWeek* called the deal "a major breakthrough in China's banking reforms." Almost immediately, Moody's, the international credit rating agency, revised SDB's outlook from stable to positive. "Moody's believes that the acquisition will have positive impact on SDB in the long run," the agency wrote. "Because of the active participation of Newbridge Capital, it is possible SDB will become a stronger and more outstanding bank."

Well, I thought, that was the whole idea.

Part II

Fight

Chapter 7

Tug of War

After the announcement, the board of SDB met for two and a half hours on Sunday, September 29, 2002, and approved the agreement empowering Newbridge to exercise control over credit decisions.

Without knowing whether the top leaders of China's government would allow the deal to proceed, we had held off launching a full-scale due diligence process—a thorough financial checkup for SDB. These were massive, expensive operations, involving teams of lawyers, bankers, and consultants and often costing millions of dollars. Now that we had received the green light, though, we assembled a large team to start doing the necessary digging. We needed to learn all we could about the operations and the true financial health of SDB. It would be a months-long process. A small team from Newbridge Capital would work with advisors from the investment bank Morgan Stanley, the auditing firm PwC, lawyers at Cleary Gottlieb and Fangda (the Chinese law firm), the consulting firm A. T. Kearney (ATK), and others. On its side, SDB would be assisted by Salomon Smith Barney.

All these institutions and individuals would be tasked with combing through the bank's accounts and financial records, focused on a few core questions: What did SDB's loan portfolio look like? How many nonperforming loans were on its books? What were the potential losses from the bad loans? And what was the commercial potential for the bank?

The due diligence work kicked off on the same day that the SDB board gave the deal its blessing. Meanwhile, we appointed our three-person transition control committee (TCC), headed by Michael Yahng. Yahng was a veteran banker and chairman of the Risk Management Society of Hong Kong, an organization that included all the city's major banks. With a bachelor's degree from Yale and an MBA from Harvard, Michael was an American but spoke fluent Chinese. He was mild in temperament and soft-spoken. We felt he was ideally suited for the job and gave him the formal title of interim CEO.

Zhou Lin called an all-managers meeting at SDB's headquarters on September 30. Standing in front of the gathering, Zhou spoke to the group first, providing the background of the proposed transaction. Then he introduced me and invited me to speak. I talked about Newbridge and what we intended to do. There was a sense of excitement among the staff—as well as angst—in anticipation of major changes. I spent half an hour with the core team the bank had assembled to assist us in our diligence process. They asked some good questions. It gave me great comfort that the bank was prepared to be transparent with us.

Yahng and his team quickly ingratiated themselves with the staff of the bank. They welcomed him as an individual, but more than that, the SDB staff was clearly keen for a foreign investor. They all knew that SDB was a troubled bank, and a foreign takeover would give it a new lease on life. While the due diligence work got underway, Yahng and his advisors started their work as what could best be described as credit gatekeepers. The TCC would update the Newbridge team a couple of times a week, so we would know what was going on. .

★ ★ ★

Until this moment, my interlocutor had been Zhou Lin, and that direct line for communication had worked well. As president of SDB, he was supposed to report to the chairman, a man I had never met, probably because he was about to retire. A few months after our conversations started, the Shenzhen government appointed a new chairman for the bank. I met with the new chairman once for dinner in Hong Kong and never saw him again. It seemed clear to us that Zhou was running the show—leading all the negotiations, being the interface between Newbridge and the city government, obtaining support from the SDB board, and handling all government approvals, from the local level to the provincial level and all the way to Beijing. He seemed to have boundless energy and wide reach. I keenly felt that the deal would not be possible without him.

But after the public announcement, this arrangement changed. The city government decided to designate a "chief negotiator" to hammer out details of the deal with us. His name was Wang Ji. Au Ngai and I met with him for the first time in mid-October in an office at SDB headquarters. Wang was in his late 40s and looked younger, thanks to his slim build and fast-moving mannerisms. He had a pair of lively eyes that betrayed his emotions. He spoke rapidly, rarely pausing for a break.

Wang had been the president of the Shenzhen Commercial Bank, a city bank also controlled by the city government. Whereas SDB was a national bank with a branch network covering the entire country, a city bank was licensed only to conduct business within the boundaries of the city in which it was located.

At the first meeting, we immediately got down to business. We spent several hours negotiating on one issue: the price we would pay for the shares. We suggested that we wait until the due diligence was completed so that we would have that crucial piece of data, the adjusted NAV. But Wang wanted to fix the price immediately. He also suggested that he could influence how the adjusted NAV would be determined, implying that waiting would not do us any good.

We found his insinuation of manipulating the NAV troubling. The work of determining the NAV figure was being conducted by PwC, the accounting firm SDB had engaged at our recommendation. Wang's suggestion worried me, first and foremost because it would entail an end run around PwC. I only grew more concerned as we picked up rumors that people claiming to be members of Wang's team were snooping around in the bank, raising questions about this and that. If the bank's staff were given instructions or otherwise pressured to be less transparent with us and with PwC, we indeed may not have been able to get to the financial truth.

Despite Wang being the new face of the negotiations, we knew that Zhou was, behind the scenes, most influential with ultimate decision makers—the top leaders of the city government. Zhou told me that Vice Mayor Song wanted a price set at 1.5 times reported NAV, which would have been at the high end of the range we had committed to pay if the adjusted NAV was zero or negative. But the adjustment had not been completed, and until then I was not ready to concede to paying the top of the range. The next day, Zhou came back to say that the vice mayor was insisting that unless we agreed to the 1.5 number, the city govern-ment would terminate the transaction. I was surprised, but I said we had a binding agreement to consummate the deal and we had to follow the agreement. I was not in a position to agree to 1.5 times reported NAV.

Zhou pleaded for us to accept the vice mayor's asking price, but I did not budge. I thought the vice mayor's threat to walk away was a bluff. Our SDB transaction had been approved by so many layers of the government, all the way to the State Council, and there had been so much publicity already that I did not believe the government could simply walk away.

In retrospect, it's clear that had I agreed to the vice mayor's price then and there, we would have saved a lot of trouble. But my reason involved a basic principle: We had not completed our due diligence. I also hoped that we could get a better price in view of the poor asset quality of the bank. I sensed that my colleague Au would have preferred for us to simply accept the asking price and not quibble over the difference, although he did not explicitly urge me to do so. He would be proved right.

In November, both PwC and Morgan Stanley completed their initial assessment of the asset quality of the bank—a key piece of the puzzle. The picture was not pretty, to say the least. SDB's own annual report for 2001 showed a nonperforming loan ratio of almost 15%, which was already an astoundingly bad figure, but PwC's preliminary conclusion was that the actual ratio was twice as bad, or as much as 30%. This meant that the bank may not have been able to collect interest and principal on 30% of its loans. The bank was technically insolvent. In fact, because like most banks, its equity capital, or the net asset value, usually covered only 5% of its outstanding loan book, it was insolvent many times over.

This was profoundly troubling. Morgan Stanley came to a similar conclusion, suggesting that SDB would require about $1 billion in new funds to be properly recapitalized. That was twice as much as the planned stock rights issue was expected to raise.

If there was a silver lining in Morgan Stanley's analysis, it was that its project team still thought the deal was worth pursuing. Their analysis was presented to us by Harrison Young, a managing director. We knew him well from the days of the Korea First Bank transaction in which Morgan

Stanley, whose team, led by Young, advised the Korean government in its negotiations with Newbridge. He said that the economics of the investment would still work, as he believed that the value we would create by turning around SDB, minus the capital needed to recapitalize the bank, would provide us with adequate returns on our investment.

Hypothetically, if someone invested $1 billion to fully recapitalize the bank and if the bank was traded, say, at three times its tangible NAV (which would be conservative for Chinese banks at that time), then the market value would be $3 billion. And if the investors owned 50% of the bank for their $1 billion, then the investment would be worth $1.5 billion almost immediately, a gain of 50%. That was the kind of math that Young presented to us.

The catch here was that Newbridge would not want to put money into the bank at the valuation for the rights issue. Even at a discount to the market price, which was at about five times reported NAV, the rights issue would be too expensive for us—not even the cleanest banks outside China were traded anywhere close to that level. If the bank issued new shares through the rights issue, those existing shareholders not buying the new shares would be diluted. We would not mind being diluted because the dilution would be modest at such a high valuation. Only in the unlikely event that we were called upon to rescue the bank by injecting our own capital could we conceivably put in, say, $1 billion for a large percentage of shares.

Therefore, Morgan Stanley's model was based on a critical assumption: that we would be allowed by regulators to operate the bank *without* adequate capital. The rights issue would raise only about $500 million, not enough to fully recapitalize the bank. The banking regulator would therefore have to allow SDB to operate for a long period of time without reaching the capital adequacy ratio required by regulation. There were precedents for what was known as regulatory forbearance in other countries, meaning that regulators would refrain or forbear from enforcing their own regulations in this regard. For example, Bank of America had benefited from such forbearance in the 1980s as it struggled to recover from its loan losses in Latin America and had taken time to replenish its capital base. But there was no telling if the Chinese regulators would do the same once we took over control of the bank. That would be one more hurdle to cross.

We knew that success would depend ultimately on how fast we could grow the bank. Morgan Stanley, together with the consultants from ATK, developed working assumptions and financial projections for SDB for the next five to seven years.

The Newbridge team held a conference call with ATK and Morgan Stanley on November 28 to review those working assumptions. It became clear to me just a few minutes into the call that their projections were unrealistic, for a number of reasons.

So far in 2002, SDB had been growing its loan book at an annualized rate of about 80% (!), which of course was an astounding number—and was probably being achieved by making loans of dubious quality. ATK's model projected a growth rate of 11% next year. We concurred that the growth rate should slow down as we tightened risk control, but we thought ATK's plan was too timid. We thought we could increase the rate at which the bank collected deposits to more than 20% per year, and typically, the lending activity of a bank had to keep pace with the deposit-taking volume.

More importantly, ATK proposed that we shift the future loan mix away from retail customers—mostly mortgage loans—and focus on corporate loans to businesses. That would significantly deviate from our plan to change SDB into a retail bank, because retail loans entailed higher interest and, being more diversified, lower risks. At the time, SDB's retail loan exposure was tiny, so a strategic shift in the direction of retail was called for. Of course the shift could only be achieved gradually.

After two hours of discussion, I asked the ATK team to set a target of a loan mix of no more than 50% for corporates with the rest leaning gradually toward retail by the end of 2005, or in three years.

After some grumbling as people debated the feasibility of achieving such targets, we reached consensus around my suggestions and we all went back to do further work and analysis. We resumed the conference call the next day to discuss the revised assumptions. In the end, everyone thought the projections—achieving 20% annual growth rate on the back of building a retail banking franchise—were aggressive but feasible.

One thing was clear: There were very limited ways to replenish SDB's capital because capital raising required complicated approval processes. Banks worldwide were supposed to follow the rules set by the Bank of International Settlements (BIS) based in Basel, Switzerland. At the time, the BIS required that banks have capital representing no less than 8% of their assets weighted by their risks. This was known as the capital adequacy ratio or BIS ratio. Even with the planned rights issue and projected earnings growth, SDB was not expected to achieve capital adequacy for a number of years. We pinned our hope on regulatory forbearance. Many banks in China probably did not meet minimum capital adequacy requirements, and the central bank knew it to be a systemic problem, hence the glaring need for banking reforms.

★ ★ ★

In recent weeks we had met with Wang Ji, the chief negotiator, a number of additional times, but we had a hard time making any headway with him. Our conversations suggested he was unaware of the history of discussions we had had with Zhou Lin and others in the city government. I had no idea why he was kept in the dark by the other side. Perhaps the city government or Zhou had considered the chief negotiator only a title, his place in the talks a façade, or perhaps they didn't want him to be biased by earlier discussions and hoped that he could strike a better deal. We had no way of knowing. In any case, it was rather frustrating to come a long way only to find ourselves at the starting point of a journey. In our negotiations, I would refer to some understanding we had achieved with Zhou and other Shenzhen officials, and Wang would dismiss it out of hand, making demands that I considered off-base or outlandish. In meetings, we often talked across each other without achieving any results or making any progress. It was as if we hadn't had any of the difficult discussions with Zhou already. I sometimes sensed Wang Ji was as frustrated as I was.

★ ★ ★

Shenzhen had numerous good restaurants, some with private rooms. It was in one of those that on November 4, Au Ngai and I found ourselves in the company of Vice Mayor Song Hai and Zhou for dinner.

The dinner was pleasant, as usual. Au and I reported to Song the progress, or lack thereof, of our work. We affirmed our commitment to the deal, although we were frustrated with not being able to reach agreement or even narrow the gap on pricing with the chief negotiator. The

vice mayor reiterated the commitment on the part of the government to
see the deal through. He then said he had heard our complaints about not
being able to make progress with Wang Ji. He told me the government
had decided to replace Wang as the chief negotiator.

It was not until the end of November that we learned that Wang's
replacement would be Xiao Suining, president of the Shenzhen branch of
Bank of Communications, the fifth-largest bank in the country.

The concept of a *branch* in China's banking system is different from
and much bigger than in most other countries. In China, a banking branch
is like the regional headquarters of a national bank in the United States
where numerous branches on the West Coast may report to a regional
headquarters located in Los Angeles. In China, such a regional headquarters is called a *branch*, under which there are many sub-branches, which
are similar to branches in the U.S. context. Mr. Xiao had reporting to him
some 60 sub-branches in the Shenzhen area, which had a population of
about nine million in 2002. In the Chinese system, Xiao ranked the same
as a vice mayor of Shenzhen, much more senior than Wang, a fact that
signaled the importance the government had attached to this deal.

Okay, then, I thought, we have a new interlocutor. We wanted to
meet with him immediately, to get a new and more productive track
going, only to learn a few days later that Xiao had checked into a hospital in Guangzhou, the capital of Guangdong Province about 75 miles
(120 kilometers) northwest of Hong Kong.

On Saturday, November 30, Au and I took the train from Hong
Kong to Guangzhou to visit with Xiao. Once the train crossed from
Hong Kong into China, the view changed, tall residential buildings like
stacked-up matchboxes giving way to rice paddy fields that stretched to
the horizon. Tall buildings only began to reappear when the train pulled
into the suburbs of Guangzhou.

The Guangzhou train station was packed with people. We took a
taxi, which negotiated its way through the busy streets filled with vehicles,
past sidewalks crowded with pedestrians. It was a clear autumn day, and
the tropical heat of summer had long dissipated. Soon, we arrived at the
Southern Hospital, affiliated with the Southern Medical University, one
of the best medical schools in the country.

So it was that our introductory meeting with Xiao Suining was held
in a hospital. The man lay in bed, with his back propped up and his wife

at his side. Xiao was in his mid-50s, and perhaps five and a half feet tall, although it was hard to tell as he was not able to get out of his bed. He looked fit but somewhat haggard and exhausted, recovering from what doctors said had been a complicated surgery. He shook our hands with a serious look on his face, but once the conversation got going, he became animated and engaging. Given his condition, we kept conversation general, about his health and broader banking issues, and steered clear of anything about the deal. But I had at least a sense that he was well versed in banking and was engaging, a man we possibly could work with.

A few days later, after Xiao had checked out of the hospital and returned to Shenzhen, we met again. Now Xiao claimed he had received no specific instructions from the city government to negotiate with us. He said the price we had discussed with his predecessors was within reason, but that without authorization from the government, he could not discuss terms with us. I did what I often did at such moments—turned to Zhou Lin for an explanation. Zhou said that Vice Mayor Song, who had been in charge of the SDB deal on behalf of the city government, was promoted and would soon become vice governor of Guangdong Province. Song was probably distracted by his pending move and new job. Xiao, he said, was a prudent man. He would not engage with us until he had specific instructions from the government.

So while we waited for Xiao, my colleagues and I took to traveling every day from Hong Kong to Shenzhen to work with the teams on the due diligence front and to speak with the relevant representatives of the city to keep moving things along.

On December 13, Zhou brought me and some of my colleagues to meet with the mayor of Shenzhen, Yu Youjun, over a dinner in the dining room of a government guesthouse. The mayor looked more like a scholar than a bureaucrat, with pale skin and soft mannerisms. I knew he was 49 years old, although he looked younger. It had been about eight months since our first meeting with Zhou to talk about SDB. During this time, we had met with a number of senior government officials of the city, including the party chief and a vice mayor, yet this was the first time we met with the mayor.

Mayor Yu was polite but not particularly warm or engaging. Nor was he inquisitive about Newbridge Capital or what we thought about the SDB deal. He did express his general support for the transaction, but I couldn't figure out whether he had been well briefed, so he had no more questions, or he just wasn't particularly interested. He came across as being easygoing but somewhat aloof. I had to search for topics to talk about. I felt that the dinner meeting was mainly a formality, and that the mayor was not much involved in the decision making on the SDB sale. It was much later that I learned I was wrong.

We finally sat down to formally negotiate with Xiao Suining on Thursday, December 19. Now recovered from his surgery, he looked in good health, and while he seemed a bit too thin, his handshake was firm. After exchanging pleasantries, he first produced a piece of paper and handed it to me. It was an authorization letter, on the official letterhead of the city government, that empowered Mr. Xiao Suining with full authority as the chief negotiator for the SDB transaction. The letter was issued by the city government and stamped with the official seal in bright red. Wang Ji, the previous chief negotiator, had never showed us any credentials. We would never doubt that either Wang or Xiao was authorized. But the fact that Xiao would not have met with us without this proof of credentials spoke volumes about his cautious style. He wanted neither side to question his authority and he wanted everything to be proper. His style was in sharp contrast with the almost casual style of the previous representatives of the bank and the government we had met to date.

My colleague Ricky Lau joined the meeting, as did a banker from Salomon, representing the sell side. We spent about four and a half hours together, but we got nowhere. In fact, the gap between us had widened significantly. A month before, the vice mayor had insisted on 1.5 times the reported NAV, the top of our agreed range. Now Xiao blew through that range and demanded 2.3 times NAV. He presented this as a compromise, saying he thought Shenzhen should have insisted on a multiple of 2.8. I was quite annoyed by this, and I told him so. I said the city government lacked sincerity and good faith. But he smiled and asked me to be patient, saying that he had to take one little step at a time. I figured that to him, this was all part of a necessary ritual of shadow boxing. But I was growing increasingly concerned that this process would take too long, and with each passing day and each new obstacle, I felt a greater risk that the deal would fall apart. There was little we could do, other than negotiate and continue to play the game together with them.

Chapter 8

Behind the Scenes

W hile we worked on all fronts to move the deal forward, events were unfolding that were well beyond our control. 2003 would bring a major reshuffle of the Chinese government, a once-every-10-years change in the top leadership. In March 2003, President Jiang Zemin and Prime Minister Zhu Rongji retired. Hu Jintao was elevated to the presidency and Wen Jiabao, who had been vice premier under Zhu Rongji, became the prime minister. Wen had given his in-principle consent to our SDB deal in 2002, along with Zhu.

The reshuffle of cabinet members had begun even sooner, toward the end of 2002. And it brought two key players into the equation who would drive changes in China's banking industry for the next decade and beyond.

In December, Zhou Xiaochuan became governor of the PBOC, the central bank. Zhou was the best trained and most experienced central banker the People's Republic of China had ever had. He had studied chemical engineering in college and obtained a PhD in automation and systems engineering from China's elite Tsinghua University. He had served as deputy president of the Bank of China, one level below the CEO, and also as the deputy governor of the PBOC while simultaneously heading up China's State Administration of Foreign Exchange, also known as SAFE, which manages China's foreign currency reserves. He later was chairman of China Construction Bank, and, most recently, chairman of the China Securities Regulatory Commission. It was quite a resume. Nobody else had such a diverse background in the Chinese financial system. He was also a renowned reformist, which I took as an excellent sign that the country was serious about taking on major banking reforms. In any other context, the sale of SDB to an American investment group would have been all but inconceivable.

Tall and trim, Zhou had the build of an athlete but the mannerisms of a scholar. I had known him for many years, since my days at J.P. Morgan. The first time we shook hands with each other, he smiled and said, "I've read your papers," referring to work I'd published while at Wharton. I was instantly drawn to him for his friendliness, intelligence, and wide range of interests and knowledge. For example, he loves music and even published a book on Broadway musicals.

I remember once talking with him about the need for businesses to build core competence and to specialize in order to succeed, which was a subject I had taught at Wharton. At that moment, we were playing badminton together. To prove my point, I remarked, "It is just like no athletes can be good at all sports." Zhou looked at me with a grin and a twinkle in his eye and said, "I am good at all sports." Everyone around us burst out laughing. Indeed, there were few sports he couldn't play well. A few years after that conversation, I saw a news report that mentioned a game of tennis in Beijing between Zhou Xiaochuan and Larry Summers, the U.S. treasury secretary, in which they jokingly bet that the winner would get to set the exchange rate between the dollar and the Chinese yuan. He won.

The second sign of a change in the wind was the appointment in January 2003 of Liu Mingkang as chairman of a new regulatory body, the China Banking Regulatory Commission (CBRC). Previously, the supervision of banks had been the purview of the central bank. Now, that function would be carved out and given to the CBRC. It was another signal of the pending overhaul in China's banking industry.

Liu, who up until that point had been chairman and CEO of Bank of China, was another strong reformer. Like me, he had been sent to the countryside to work as a farm laborer for more than 10 years during the Cultural Revolution, and subsequently for three years as a factory worker.

Then he rose through the ranks from a junior staff member at Bank of China to, improbably, the vice governor of Fujian Province (with a population of 40 million in 2020), then to become deputy governor of the central bank. Somewhere in between, he had obtained an MBA from the City University of London.

I had met Liu for the first time a year earlier. A friend of mine, a senior partner at PwC, the accounting firm, called me one day. She said that Mr. Liu Mingkang, chairman and CEO of Bank of China, wanted to meet with me. I inquired as to the purpose of the meeting, but my friend did not know. As I was traveling to Beijing on business a few days later, I was invited to join him for dinner. I had expected to see other guests, but was quite surprised to discover I was the only one there.

Pale and soft-spoken, with his hair neatly combed, Liu came across more like a professor than a banker. "Let me come straight to the point," he said, even before we had sat down. "We are restructuring Bank of China Hong Kong Group before bringing it public. I would like to invite you to be an independent director of the restructured bank."

Bank of China (Hong Kong) Limited, or BOCHK, had a long history in Hong Kong. It had been established in 1917 and had grown to become the city's second-largest banking group (after HSBC). Liu was positioning it to become the first Chinese bank to go public on the Hong Kong Stock Exchange, paving the way for other major Chinese banks to follow.

Liu was familiar with the success we had with turning around KFB, he told me, and he wanted to tap into my experience in banking to help with the restructuring and the planned initial public offering of BOCHK.

I was happy to help. I agreed to serve not only as an independent director but also as chairman of the bank's audit committee. At the time, almost nobody wanted to serve on the audit committee of a public company because of the risks it entailed. Just a few months earlier, the U.S. energy trading company Enron had imploded spectacularly, prompting the collapse of its external auditor, Arthur Andersen, a century-old accounting firm. The scandal also gave rise to significant personal liabilities for the members of its board's audit committee. I agreed to serve because I appreciated Liu's trust, and I thought I could help his bank.

In another sign of how open Liu was to fresh thinking, he had invited a number of prominent bankers and business leaders from abroad to serve as independent directors, and had given us a strong voice on the board. He even conducted board meetings in English, even though that meant some of his own colleagues required simultaneous translation.

Looking at the landscape of reform generally and these two appointments in particular, I was thrilled with the changes. I thought they heralded further reforms that might well lead to a more open and market-oriented economy.

Since I knew Governor Zhou and Chairman Liu well, I made sure to keep them informed of the progress of the SDB transaction, which now fell under their purview. Not surprisingly, they were both supportive of the transaction. They understood that it had the potential on its own to be a significant step in China's banking reforms. I also shared with each of them the results of our due diligence work to date, including the report produced by PwC. This report was probably among the first quality inside looks at a Chinese bank ever conducted by an international audit firm under the direction of a foreign investor. Neither man appeared surprised by the extent of the problems; both wanted to see transformative improvements in the bank's operations.

Our due diligence work had already yielded a good deal of information. We had long suspected that SDB's annual reports did not reflect the true conditions of the bank. Now, it was obvious to us that management had window-dressed its financials by misclassifying and under-provisioning the bank's NPLs.

Classifying and provisioning for loans is a subjective process, one that often depends on the views of the person making the assessment. One loan officer might classify a loan whose interest is still being paid by the borrower as "questionable" because the officer has judged the borrower's ability to repay the principal to be impaired. This type of classification is called "forward looking." But another officer may consider that same loan "normal," because it is still performing (i.e., being paid) with no apparent trouble. International best practices encourage banks to use the forward-looking methodology, but banking in China in the early 2000s reflected old practices. It was far more likely for a bank like SDB to cover up its problem loans by offering new, even bigger loans to the same borrower to enable them to pay back the old ones than to declare them bad loans—a Ponzi scheme, in effect. Since it would be difficult for auditors to examine each loan and borrower, they often missed such practices.

It was not surprising to us that SDB's domestic auditor would sign off on annual reports that hid more than they disclosed. We had never expected a high standard for Chinese domestic auditors. However, SDB also had engaged a foreign auditor, one of the biggest accounting firms in the world. The foreign auditing team told us during our meetings that they would stand by their stamp of approval of SDB's annual reports as well. I supposed that that team did not want to disavow its own work— but its team members became defensive when we questioned it. More surprisingly, its team leader promptly complained to Zhou Lin, accusing us of putting pressure on them to make more provisions for bad loans.

While it was in our interest that the problems of the bank be fully disclosed, I had not considered the full implications, especially in the stock market, of such disclosures. The biggest risk for any bank, of course, lies in a loss of confidence among depositors and the market, which in the extreme can trigger a bank run. Even though Zhou Lin knew the extent of his bank's problems, he became quite upset when the accounting firm told him of our "pressure." I did not realize the extent to which he was worried about the market reaction to an honest disclosure of SDB's problems. He likely also considered it a personal affront or a sneak attack by us behind his back. Although we did not know it at the time, the seed of mistrust between Zhou and ourselves had probably been planted.

★ ★ ★

As we moved into the new year of 2003, I grew worried that the Shenzhen side had slacked off in its efforts or lost some of its enthusiasm. It had become difficult to get a response to our repeated efforts to engage, and to close the pricing gap via negotiation. I suspected something was amiss. What were they up to?

By now the PwC audit was largely complete. And the news was not comforting. PwC downgraded 16% of SDB's "performing loans" to "substandard" or below, or from good loans to bad. This was in addition to the 15% or so NPL ratio given by the bank itself. Meanwhile, that magic number, the adjusted NAV, was significantly negative, worse than Zhou Lin had first suggested. The work by PwC showed that this bank really was broken. The only reason SDB was still operating was because of the abundant liquidity provided by depositors, who continued to pour money into the bank.

Under the framework agreement, we were to pay five times the adjusted NAV figure. But now that it was clear that the adjusted NAV was negative, we would obviously need a different metric. We could only consider a multiple of the nominal or reported book value, as the parties had anticipated from the beginning. With the fresh revelations of the true extent of SDB's bad loans, we were increasingly concerned that we would be overpaying for the bank. The SDB books were in such bad shape, in fact, that it seemed we would be paying a substantial price for a banking *franchise*, with no equity capital to its name.

So we had a bank whose troubles had been confirmed, and a negotiating partner that had gone silent. We really did not know what they were thinking.

In January 2003, the *21st Century Economic Herald* reported that Newbridge had run into difficulties in the SDB transaction, citing the asset quality of SDB as a major issue. Probably not coincidentally, someone affiliated with another private equity firm called to tell me that Shenzhen had approached their company to gauge interest in SDB. I was worried. Was Zhou Lin having second thoughts regarding our deal? Was he shopping around for his bank?

I continued to travel to Shenzhen almost daily, to meet with Xiao Suining and his team. On the surface, nothing seemed to have changed. On Friday, January 10, Zhou Lin invited me to address the annual meeting of SDB's employees. It was an unseasonably warm and sunny day in Shenzhen and the meeting was held in a resort hotel. The employees greeted me warmly and showed interest in what I had to say. After all, everyone knew I represented the future, as a controlling shareholder in their bank.

That afternoon, Au Ngai and I met with Xiao in his office at the Bank of Communications to continue to negotiate on pricing. Xiao indicated that Shenzhen's ask had come down to 1.65 times reported NAV, from the 2.3 multiple he had asked for some three weeks before.

Ever since we had first begun to negotiate back in June 2002, the asking price from the other side had not been based on any good reasoning that I could figure out. It seemed to simply be the highest possible price they thought they could get on any given day, or perhaps it was mere posturing to push us to bid higher. It was understandable that a seller would want as high a price as possible, while the buyer wants the opposite. But to reach a deal, the expectations of both parties had to converge. I welcomed Xiao's offer to reduce his own ask but still rejected it as unreasonable, referring to our earlier agreement with Zhou Lin to cap the figure at 1.5 times. Based on the information we had by now gathered, we thought that even 1.5 times seemed too high.

On our way back to Hong Kong that day, Au shared his thinking. He believed both parties could eventually converge on a price of one billion yuan ($120 million) for the stake, a number he thought would be acceptable to both sides. That price would translate to 1.38 times reported NAV. I agreed with him. While we had hoped for a price closer to the reported NAV, in keeping with prevailing market valuations, I thought we could live with paying a premium for the controlling stake. Our sense was that *one billion* should psychologically appeal to the other side. It was a nice-looking whole number. Au and I also shared the view that we needed to get this done quickly. We were concerned that any further delay would risk the Shenzhen side changing its mind about doing the deal with Newbridge.

In an internal conference call on January 28, our partners all agreed to the price Au and I had in mind. Blum and Bonderman urged us to make a greater effort to reach a compromise. They said that, above all, we had to maintain momentum.

For us, knee-deep in the negotiations and all the inner workings of SDB, our position seemed eminently sensible. But Bonderman provided a different perspective. The difference between Xiao's current ask of 1.65 times and our thinking of 1.38 times would be $25 million. That seemed like a lot of money and it represented a 20% premium to the one billion yuan number (roughly $120 million). But Bonderman didn't see it that way. "This deal," he said, "is either going to work or it isn't. The result will be binary." In his mind, we either would make a lot of money if we could successfully turn around this bank or we would lose money if we failed to do so. "In the scheme of things," he continued, "$25 million makes no difference." The key, he felt, was to seize the moment and get it done as quickly as possible.

Bonderman had a gift for seeing, and focusing on, the big picture. In rapid fashion he had helped us see that we had been too deeply enmeshed in the weeds and the details. We quickly saw his point—and came to his side of things. Now we wanted to get the deal done. If we had to pay 1.65 times the reported book value, so be it.

★ ★ ★

The Chinese New Year fell on February 1 that year. Typically, the public holidays that follow stretch for a week or 10 days, including weekends on both ends. My wife and I had planned to take our daughter for a ski vacation in Japan during the holiday. But then I learned that the two most senior banking regulators, Zhou Xiaochuan and Liu Mingkang, would be spending the holidays in Shenzhen.

I didn't like combining business with family time; nor, of course, did I wish to upset a family vacation. But this was a crucial moment for the SDB deal, and the two top regulators were rarely as close by and accessible. And in fact, my wife was not much of a skier, and our 10-year-old daughter was happy to vacation anywhere. There was an amusement park called Window of the World in Shenzhen with, as the name suggests, models of global landmarks laid out in miniature throughout the park. We found a hotel near Window of the World, canceled the Japan trip, and on February 2 made the much shorter trip to Shenzhen instead.

Xiao agreed to meet with me that evening in our hotel. It was 9:30 p.m.; my wife and daughter had already retired for the night. I took as a good sign the fact that he was willing to meet at such a late hour on a major holiday at a place quite far away from his home. But as it happened, the urgency had a different source, and what he brought was not good news. He told me that the situation was taking a turn for the worse. I took him to mean that the Shenzhen side was having second thoughts about the deal. But he gave no reason. He said only that he was instructed to give us an ultimatum: a final offer of two times reported NAV.

Needless to say, I was astonished and disheartened. Less than a month before, he had asked for a multiple of 1.65, and while we had made clear that that figure was too high, I was ready to hit that price. Now, Xiao was moving even further away from his own asking price. He clearly knew it was absurd. He apologized but said he had to follow orders; he was only representing the government. I could not pry from him who the decision makers were or who had given him the instruction.

I sensed that Zhou Lin was withdrawing his support. Zhou had already impressed us with his political capabilities; it seemed that he could make anything happen, either in Shenzhen or in Beijing. I also believed that he could stop the train in its tracks if he wanted. So I called Zhou and confronted him with my suspicions. He neither admitted nor denied it. But it was clear to me that he had had a change of heart. Now I thought I understood: The asking price of two times NAV was his way to break the deal, by forcing us to walk away. This epiphany hit me like a brick. My heart sank. All the hope, all the effort we had put into this, and now the possibility of losing the whole deal seemed very real.

But we also knew this much: The Shenzhen side did not have the right to simply walk away. They had signed agreements that bound them to a deal, setting a price range between 0.8 and 1.5 times reported NAV. Even in the worst case, we could accept the top of the range and then the agreement would be legally binding. At least, that's what I thought.

On Thursday, February 6, I took another brief break from my vacation and went to the Wu Zhou Guesthouse to visit with Liu Mingkang, the CBRC chairman. He was having lunch with Shenzhen mayor Yu Youjun and the city's deputy secretary general, Liu Xueqiang. When the mayor saw me, he asked how the SDB deal was going. Apparently he had no idea that anything had changed.

I told him only that we were still negotiating, still eager to get a deal concluded, but that his side was vacillating on the purchase price. He said that many central government leaders were expressing interest in the deal, and that Prime Minister Wen Jiabao's secretary had called him to inquire about the progress. They all wanted to see the transaction consummated as soon as possible. I thought to myself: *We had every intention to get the deal concluded; it's your side that's holding things up.* But I held my tongue. Clearly, I thought, Mayor Yu was not in the know.

After the mayor left with Liu Mingkang, I stayed behind to talk with Liu Xueqiang. I told him about the problems we had encountered. He said that the big decisions had to be made at the top, suggesting that Zhou Lin alone was not the ultimate decision maker. He also threw another element into the conversation. The price, he said, should be no lower than the Citi/Pudong deal. It had been reported that Citibank had bought 4% of the shares of Shanghai's Pudong Development Bank at 1.54 times reported NAV. Since we had internally decided that even 1.65 times was acceptable to us, I assured him we were willing to pay at least the Citi/ Pudong multiple for SDB.

We soon confirmed via various back channels that Zhou Lin had turned from a leading advocate for Newbridge to its main opponent, for reasons we could not fathom. Maybe the fissure had opened when I declined his plea for us to accept the 1.5 multiple requested by the vice mayor, or when the foreign auditing firm told him we wanted more conservative provisioning against bad loans. In any case, I knew his changed position was likely to jeopardize the deal, or at least make it very difficult for us. It had become increasingly clear that he wanted us to withdraw without a fight. I felt quite helpless, thinking through it all. How were we to reverse the downward spiral, with Zhou or anyone else?

But the nature of this whole process, from the beginning, was its wild rise and drops that quickly turned elation to exasperation, reversing course over a single call, meeting, or rumor.

And so it was at this moment, that we heard suddenly that Mayor Yu remained supportive of the deal. Yu was second in command in the city— the top decision maker was the party secretary, who was said to be leaning toward Zhou Lin's views—but I was pleasantly surprised to hear that the mayor was still on our side. In a complex deal like this one, you never knew from which quarters support or opposition would come.

Zhou had never given us the impression that the mayor mattered much. It was telling that Zhou hadn't arranged for us to meet with Yu until weeks after the documents had been signed, and I couldn't tell at that meeting whether the mayor's expression of support had been more perfunctory than whole hearted. Nonetheless, we were now hearing that the mayor wanted the deal done in accordance with our earlier agreements. Apparently, he thought that whatever terms the city government had committed to had to be honored. A basic principle, you might think, but it came as such welcome news at this juncture.

I next met with Xiao, the chief negotiator, after the Chinese New Year holiday. Xiao had conducted himself professionally and responded well to reason. It was not surprising that the two of us had developed a mutual respect. It was also clear, however, that his authority was limited.

Xiao came to this meeting with new instructions. He made three points. First, he wanted Newbridge to commit to keeping management control for at least five years. This request, he said, had come from the mayor. Second, he said the price should be 1.65 times stated NAV in the accounts of the bank as of September 2002. Third, if we did not accept the requests within a week, the city government would speak with other potential investors. He added, as if to inject additional urgency, that a third party had already offered to pay 2 to 2.5 times stated NAV for the same stake.

By our framework agreement with the Shenzhen government, we had exclusive rights to the deal. The government was not supposed to speak with any third parties. But our negotiations with the government had already become public knowledge, so it was not surprising others came forward to show interest.

It was only much later that we managed to piece together what had happened. There were two camps in Shenzhen's leadership. One questioned our commitment to own and manage the bank, and our willingness to accept the proposed price of 1.65. That camp advocated for handing the deal to a new bidder. The other camp, led by the mayor, had clearly wanted the government to honor the deal with Newbridge. The two camps must have reached a consensus, however tenuous, that if Newbridge accepted the conditions that Xiao was proposing now, on February 12, Shenzhen would proceed to conclude the deal.

In either case, we had come to a critical moment. Xiao's requests felt more like an ultimatum than an invitation for a counterproposal. On the one hand, I had to take the proposal seriously, and respond positively, to give Xiao and the deal's supporters the ammunition they needed to continue to support us. At the same time, I was not prepared to simply accept his proposal unless we got something in writing.

I responded carefully to his points, one by one.

"We welcome the government's support for us to have management control," I began. I told him we saw eye-to-eye on this, and that Newbridge would commit to keeping control for at least five years.

Then I told him I had documents in my possession showing that the government side had agreed to "a price range of 0.8 to 1.5 times stated NAV as of June 2002." I produced two documents and read excerpts to Xiao. One was a letter of undertaking issued by Newbridge that referred to the price range. The other was a submission by SDB to the central bank on August 6, 2002.

Lastly, I reminded Xiao that our promised exclusivity was open-ended. I read him the relevant paragraph in the June 21 framework agreement. That language had made it clear: If the Shenzhen government spoke with any other potential investors it would be in breach of a legally binding agreement.

It was a brief and lawyer-like presentation. Before leaving, I told Xiao that as a show of maximum good faith, we would offer to do the deal at the very upper limit of the agreed price range (i.e., at 1.5 times the June 2002 reported book value). Finally, I said that if Shenzhen continued to insist on 1.65 times the September NAV, it would have to put that demand in writing in order for us to take it seriously.

"Good heavens!" Xiao exclaimed, once I had finished. Clearly he had been astonished by the documents, and the fact that SDB had made the submission to the central bank. Like his predecessor, Wang Ji, Xiao clearly had not been given the background of the various discussions between Newbridge, SDB, and the Shenzhen government. Xiao searched but could not find a copy of the document on his side. I faxed it to him the next day.

For all the ups and downs, and twists along the way, I had to admit Xiao was professional. I took his surprise to be genuine, and I believed he would not have suggested something had he known he was in clear breach of previous agreements. He had staked out his positions without having any knowledge of prior arrangements. Seeing the documents must have shocked him.

Chapter 9

Roller Coaster

While I was busy with Xiao Suining, the chief negotiator in Shenzhen, my colleague Au Ngai traveled to Taipei on February 12, hot on the trail of Zhou Lin, whom we had heard was there on a business trip. Au called to tell me that he had learned, almost as soon as he had landed, that Zhou was meeting with senior people from Chinatrust Commercial Bank, a Taiwanese institution, to buy the shares of SDB. So we really did have competition. Zhou had shifted his support from Newbridge to Chinatrust. We could only guess what had motivated him to make the switch—and we started to get the feeling that we were in for a ride that had more excitement than we had hoped for.

It was understandable that Chinatrust would seek control of a national bank in China. Since the early 1990s, many Taiwanese firms had invested billions of dollars on the mainland. For Taiwanese firms, the mainland market was both enormous—with its population of more than a billion people—and convenient, as Chinese on both sides of the Taiwan Strait speak Chinese and share a similar culture. Taiwan's own banking market was overcrowded, with more than 50 banks serving a population of about 20 million. As Taiwanese firms built their businesses on the mainland, Taiwanese banks wanted to follow their clients, in addition to tapping into the vast potential there.

In addition to the entrance of Chinatrust was even more unsettling news. Back in Shenzhen, we heard that we were being discredited by malicious rumors within the city government, stories suggesting we planned to flip the SDB stake to Chinatrust after we purchased it. It turned out that was why Mayor Yu, dubious of the rumor, had wanted our investment locked in for five years. We hoped we had put the rumor to rest when we agreed unequivocally and without hesitation to commit Newbridge to keeping management control of SDB for at least that long. In any event, we had always thought it would take some years to turn around SDB, and doing so was our ultimate goal—and our means of making the most of our investment.

By then, both the Chinese and foreign press had begun to speculate that the negotiations for SDB were on the rocks. On February 14, Xiao granted an interview to a Chinese publication in response to a *Wall Street Journal* article about the deal's roadblocks. Xiao said, helpfully, that negotiations were still continuing "in a healthy, friendly, and normal manner." He and I also agreed to issue a joint statement along the same lines.

Meanwhile, Xiao had taken up my challenge to send over his new proposal, in writing, that priced the shares at 1.65 times the reported NAV of September 2002. He faxed me a letter doing just that, but not without complicating matters further. The letter said Newbridge had to accept the terms unconditionally within seven days, but that the Shenzhen side would not be bound by the price. I called Xiao to demand an explanation. There was none.

I thought this unilateralism was unfair and unheard of, especially in view of the history of our discussions and various documents we had signed. He said he had not seen the documents I was telling him about. I proposed that we meet in Shenzhen the next day when I would bring him a copy of all the documents in my possession. It was getting more and more difficult to keep this deal on the rails.

The following day, I drove to Shenzhen. To save time and avoid the prying eyes of others, Xiao came to meet me and was waiting for me in the parking lot on his side of the border control. Unable to find a quiet restaurant or hotel nearby, we met unceremoniously in my minivan. The parking lot outside the border control hall was quite immense and mostly empty, surrounded by buildings on three sides and a road without much traffic. The weather was nice; the summer heat had dissipated but the air was still warm. We kept the vehicle door open to let in the breeze.

I showed him the full set of documents we had signed the previous year, including the framework agreement, the net asset value adjustment agreement, and the PwC engagement agreement. I also showed him our pricing undertaking and the submission by SDB to the central bank, which was marked "top secret."

Over the years I had learned to temper my emotions at such moments. During all the ups and downs of our negotiations to acquire KFB, I had let my frustration show on occasion. Our Lehman advisor, Michael O'Hanlon, had pulled me aside once to remind me, "None of this is personal. It's business." I had taken his counsel to heart. Here, with Xiao, was a moment to take a deep breath, keep my composure, and hide my worries.

"Now you ask us to accept a higher price, but your side will not commit to it," I said to him. "If we did, all the previous agreements would become invalid. If you later change your mind, we would have nothing to go back on."

As Xiao flipped through the documents, he kept uttering, "Good Heavens, Good Heavens," in a muffled voice. Xiao was clearly shocked. He realized now that he had been blindsided by his colleagues, and that all the documents gave us an airtight contract.

I also realized that Zhou Lin was probably the only person on the Shenzhen side who knew about all these documents, and that he had probably never shared the full portfolio with others in the Shenzhen leadership. Now Xiao was worried on behalf of the government. He knew we could resort to legal recourse if the government tried to walk away from the deal or change its terms. He mumbled, more to himself than to me, something to the effect that we had some "nuclear bombs" in our possession. He did not know what else to say, other than to remind me that he had to take directions from the city government. But he asked that I respond formally to his letter with whatever we wanted to say, and to be sure to make reference to the contractual terms.

I went back to Hong Kong and wrote two letters. One was a cover note to Xiao, attaching a letter my partner, Dan Carroll, had written to Chinatrust a couple of days earlier, in which Carroll warned that we would be prepared to take legal action if the Taiwanese bank continued to interfere with our contract with Shenzhen. The letter read, in part:

> *Please be advised that the continuation of your discussions with the Selling Shareholders is detrimental to the completion of our investment in SDB and contravenes our exclusivity arrangements, which we will use all available legal means to defend. This may include, but not be limited to, an action for tortious interference brought in the State of Texas.*

"Tortious interference" is a claim under U.S. law that a defendant intentionally caused a third party to breach its contract with the claimant, resulting in economic damages. Carroll's letter warned Chinatrust to cease and desist from such interference, or else Newbridge was prepared to sue. We had decided to take action in the United States, and specifically in Texas, where Newbridge and TPG had operations, because Chinatrust also held assets in the United States and because the Texan courts were known to be sympathetic to injured parties in such cases.

By attaching the letter to Chinatrust, we wanted to demonstrate our resolve, and also the fact that, contrary to the malicious rumor, we had no secret deal with it to flip SDB.

The second letter was a formal message from me to Xiao, dated February 17. I wrote that we had a series of legally binding agreements that would bind both parties to a price range of 0.8 to 1.5 times the reported NAV of June 2002. However, I emphasized that we would accept the top of the range (i.e., 1.5 times).

> *To show our maximum good faith, in consideration of certain concerns of your side, and in order for us to conclude this transaction as soon as possible, I wish to repeat my offer made to you during our last meeting: Newbridge is prepared to accept the very upper limit of the price range agreed by both parties, i.e., 1.5 times net book value of Shenzhen Development Bank reflected in its June 30, 2002 financials.*

After all these latest turns, we were at a stalemate. Although the Shenzhen side could not deny the existence or substance of these documents, their officials were sticking to their more recent offer—the

1.65 multiple—without committing themselves to the price. I knew that some officials in Shenzhen were supportive of the deal, but if we stayed stuck on the price, it would only provide more ammunition for Zhou Lin. The more this matter dragged on, the more people might be persuaded to side with our opposition and the more our position would be undermined, despite the legal documents in our possession. I was also mindful of what Blum and Bonderman had said: We should not get stuck on a $25 million difference in price.

After some internal discussion among our partners, we decided that we should send a letter to Shenzhen to formally accept its offer of 1.65 times reported NAV as of September 30, 2002, and to make it binding on both parties.

My letter, dated February 19, 2003, was addressed to the party secretary and the mayor.

> *Upon acceptance by Newbridge, this price (1.65 times NAV) would become final and signify the end of all price negotiations as well as the closing of the transaction.... After having an urgent discussion and a serious debate on this matter, Newbridge's investment review committee has decided to conclude the transaction by accepting the latest proposal and the purchase price offered by the city government under the framework of the Agreement on Share Transfer and Acquisition as well as other supplementary agreements signed by both parties in June 2002. Nevertheless, Newbridge's decision to accept the latest proposal by the city government should not be viewed as us forgoing the validity and rights of all the agreements signed by both parties previously but rather as a continuation of our sincerity to fully cooperate with the city government.*

The key phrase was "the end of all price negotiations." We were hoping my letter would break the deadlock and move the deal forward.

By accepting Shenzhen's price, we were helping the deal's supporters on the Shenzhen side. There was now no reason for Shenzhen to dither further. I spoke with Xiao after the letter had been sent. He remained noncommittal, willing only to say things were "moving in a good direction."

★ ★ ★

The ball was now back in the Shenzhen government's court. In addition to the party secretary and the mayor, those participating in the decision making included two or three vice mayors, the deputy secretary general, Zhou Lin, and Xiao Suining. Ultimately, the buck would stop with the party secretary. But the mayor had a big voice too. I learned later that they had debated among themselves and eventually the majority voted to continue the transaction with Newbridge.

Xiao communicated the news to us, and we agreed to release a public statement on February 25 that the Newbridge–SDB deal was "making progress." We were encouraged, and we wanted to put out a statement to quell rumors to the contrary. Zhou seemed back on board, and even added a sentence in the draft statement that the bank had performed well since the beginning of the year. The release was immediately picked up by the press and several reporters called me. I declined to make any further comments.

With the price finally settled, we resumed formal negotiations on other details. On February 26, I drove to Shenzhen after lunch, along with Dan Carroll, Ricky Lau, and Fifi Chan, a new member of the team. Xiao discussed a few open issues, including the number of board seats Shenzhen could secure for us, the precise number of shares the various government-controlled entities would sell to Newbridge, and the timing of the rights issue to raise capital for the bank.

At the meeting, we agreed that Shenzhen would help secure Newbridge's control of the board with a series of measures. It would allow us to appoint a majority of board seats and provide us with an anti-dilution right so that the SDB shares we would own represented at least 17.9% of the total. It would also assist the bank to obtain the consent of the CSRC, China's securities regulator, for the SDB rights issue, which we still needed in order to raise fresh capital prior to the deal's closing. We also agreed that the entities controlled by the government would put all their SDB shares into an escrow account immediately, pending delivery of those shares to us. By the end of the day, I felt we were getting close to the finish line.

One week later, on March 2, our lawyers sent their counterparts in Shenzhen a detailed term sheet, outlining the basic terms and conditions of the agreement that had been under discussion for three months. The idea now was that the contents would be incorporated into the final sale and purchase agreement for SDB shares.

The Shenzhen lawyers sent back their marked-up version only three days later. After three or four more rounds of exchanges, it seemed, again, that we were about to finalize the document.

Then, unexpectedly, came another swerve of the roller coaster. We had had an understanding from the beginning that Newbridge would have control of the SDB board of directors. Indeed, Xiao had pledged as much just a few days earlier. SDB had a 17-member board; control would require that we appoint at least nine members. But now Zhou Lin balked. He would not agree to give us more than eight seats, and since he was to be board chairman, he would hold the swing vote.

We insisted, again, that we had to be able to control the board. Three months earlier, we would not have doubted Zhou's desire to work with us or his motives. But by now it was obvious that he was vying for control and placing fresh obstacles in our path.

After some back and forth, the Shenzhen side agreed to give us unequivocal control. It turned out that this had never been an issue with the government; the matter had arisen only because Zhou had a different agenda. But even he could not override the wishes of the Shenzhen leadership, which had no issue with giving the foreign investor unambiguous control over the board.

We were now back on track; the obstacle before us that threatened to derail the deal had been removed. One more time, it seemed the end was near. And then we heard yet another piece of disturbing news.

Through back channels, we learned that Jeffrey Koo Jr., chairman of Chinatrust, our Taiwanese competitor, paid a visit to Shenzhen's party secretary, vice secretary general, and others. The meeting was undoubtedly facilitated by Zhou Lin. During the meeting, Koo had offered to pay the government 2.5 times stated NAV for the SDB controlling stake (compared with our 1.65 times). We were also told that Koo had disparaged Newbridge, calling us short-term investors and claiming that Dick Blum and David Bonderman had personally told him that Newbridge did not really care about this deal. These, of course, were lies. But it didn't matter. The party secretary now wanted the deal to go to Chinatrust.

We could not let the lies stand, and once again we had to protect the deal.

We countered by having Bonderman and Blum send a letter to the party secretary and the mayor to reiterate our commitment, and our promise to own the bank for the long term. We were not in it for a quick win.

The letter seemed to have done the trick. By March 19, the new term sheet was near completion. In a typical process at this stage, the parties sign the term sheet to lock up the terms agreed between them, and then proceed to drafting the final documents. But Xiao still had to wait for authorization.

Xiao, however, felt confident. Authorization was only a matter of formality, he told us. He suggested that we not waste time waiting, but immediately move to the next stage of drafting the final transaction documents.

★ ★ ★

The next day, March 20, brought word that the United States had invaded Iraq. The world was suddenly in turmoil—global markets, too—but it was business as usual for us. Xiao and his team were working intensively with us to finalize documents. On March 25, lawyers representing Newbridge sent the first draft to the government side for their review.

Zhou, meanwhile, had not ceased his efforts to derail our deal behind the scenes. We were worried, knowing he was likely to be as effective at blocking our path as he had been at pulling strings to help us earlier. We raced toward the finish line with Xiao's team, hoping to preempt any last attempts to sabotage the transaction.

Our fears proved well-founded. We soon heard, through back channels, about yet another high-level meeting among officials from Shenzhen's government on Sunday, March 30. We were not privy to what exactly had transpired during the meeting. But from what we could piece together from various sources, there had been a rift among the leaders regarding the Newbridge deal. Yu Youjun, the mayor, had set the tone by stating that Shenzhen needed to move forward quickly with the SDB transaction. Zhou Lin had argued strenuously in favor of Chinatrust, on the grounds that the Taiwanese bank would pay a higher price. Xiao Suining rebuked him (mildly, I was sure, as he was never confrontational by personality) saying that the auction had long been over, agreements had already been signed with Newbridge, and the city government would subject itself to litigation if it walked away from the deal now. Xiao's position was supported by the mayor. A deal was a deal, he said. To this, Zhou countered that he was willing to take the consequences if Newbridge sued the city government for breach of contract, because he could claim that whatever he had signed had not been authorized.

But others in the room all knew that Newbridge representatives had met with the party secretary, the mayor, the vice mayor, the vice secretary general, and other senior city officials in connection with the SDB transaction, and that the vice secretary general had been present at the signing of the framework agreement. It would be preposterous to argue that Zhou had signed the agreements without the knowledge and authorization of the top leadership. In the end, despite Zhou's opposition, the meeting resolved to close the deal with Newbridge.

★ ★ ★

Our patience tested, our resolve intact, we moved to what would surely be the final stages—the last mile.

Now that the city government had formally made a decision to go to the finish line with Newbridge, we and our lawyers worked closely with Xiao's team and their legal advisors to put finishing touches on the term sheet and to make progress on the final agreement. By mid-April, the negotiation teams on both sides produced execution copies of the term sheet. We were ready to sign it, but Xiao said it would take a week or so for him to go through the government's internal process again before his side could execute it.

Meanwhile, in the background, Zhou did not relent. He continued to push for Chinatrust, which was proving to be a very aggressive suitor. Chinatrust had even offered the city government to pay the legal expenses and damages if Newbridge really did bring a lawsuit against the government for breach of contract.

At this time the Shenzhen government, along with every local government in China, was going through a change in leadership. Song Hai, the vice mayor in charge of financial affairs, who had been supportive of the deal, was leaving to assume a new position as vice governor of Guangdong Province. A man named Chen Yingchun took his place. Chen was new to the job and did not know the full history of the SDB transaction, but what he did know was that it was a high-profile, controversial deal, and he wanted to proceed cautiously. I was worried that we would be starting all over again. Soon, Chen invited me and Au Ngai to a dinner on April 21.

The dinner was to be held in the canteen of the government building. Au Ngai and I first met with Chen in his office. Chen greeted us warmly. He was tall and somewhat stocky. He wore a pair of dark-rimmed glasses and spoke in a rather loud voice. When he noticed that I was staring at a model artillery gun on a table in his office, he told me he that held the rank of colonel in the reserve force of an artillery regiment, and that he would go for training on some weekends. The model, he told me, was of a gun he was trained to use. I found it curious; I hadn't even known that China had a military reserve. Chen was the first reserve officer I had met. To me he looked more like a bureaucrat than a soldier, sitting there in his office, behind those glasses. Still, I reflected, at least that explained the loud voice.

Whatever anxieties we had about the new leadership were washed away during the dinner, which was also attended by Xiao and a new vice secretary general, Yu Weiliang. Chen told me in no uncertain terms that, "caution" notwithstanding, the city government remained committed to doing the deal with Newbridge.

The pace on Shenzhen's side picked up after this meeting. Xiao suggested that to save time, we forget about the formality of signing the term sheet first, and go straight to working on the formal sales and purchase agreement (SPA), which would be the final document. We agreed and had our lawyers send over a draft SPA. Soon, Xiao informed me that his team was working hard with Junhe, the law firm representing the government, as well as the government's own legal department, to mark up the draft SPA our lawyers had sent them. He and his assistants requested that we stand ready to sign the SPA by May 1, China's Labor Day, which preceded a week-long public holiday. And they followed up by sending their marked-up version on April 25.

On April 29, both parties completed the final touches to the documents. Xiao's assistant told us that he had personally delivered the final documents to Vice Mayor Chen for his sign-off. We were hoping to sign the following day.

Chapter 10

Transition to Nowhere

J ust as we thought the roller coaster was about to finally stop, it veered sharply off course.

The night before the signing date, April 29, Xiao Suining called me to say there was a change to the timetable. The vice mayor, for reasons he would not share, had postponed the date of meeting with Xiao until after the week-long holiday. Why did the vice mayor want such a long delay? What did he or the city government have in mind? We had no way of knowing, and naturally I was dreading the possibility that the government might be having second thoughts. Third or fourth thoughts, really.

Later in the day we learned that the CCP's standing committee in Shenzhen, which was the highest body of decision making, had held an ad hoc meeting that morning. One of the agenda items, again, was SDB. Obviously some officials were trying to reopen the case and reverse the previous decision. We could only guess that Zhou had managed once again to upset the process.

What we found out was even more bizarre. Chen, the new vice mayor, had suggested that the government should now simultaneously talk with Newbridge *and* Chinatrust regarding the sale of SDB's control. So much for his expressed commitment to the deal, but we all knew his commitment did not mean much without the consent from his higher-ups. We also knew he wouldn't have suggested this without someone higher up having whispered it in his ear.

It was a maddening development. The government had been working with Newbridge for more than a year and had a framework agreement with us. Now they wanted to formally invite Chinatrust into a new process? Even though Chen appeared not to take sides, his objective was quite obvious: He was advocating for Chinatrust.

His proposal must have surprised many in the audience. The decision to proceed with Newbridge had been affirmed repeatedly, most recently just one month before, in a similar meeting led by the mayor. In the wake of that decision, the government's chief negotiator had already completed the final documentation. Now Chen was pressing for a reversal, a retreat from the Newbridge transaction. Yet he presented this in such a nonchalant way, as if nothing unusual was happening.

Anyone with any knowledge of the workings of the Chinese political machine would know that the vice mayor could not have made this move without the prodding or full support of the party secretary. I figured Zhou must have persuaded the party secretary to change her support from Newbridge to Chinatrust. Not surprisingly, Chen's suggestion was immediately echoed by the party secretary, who added, no doubt for the appearance of being fair, that Newbridge should be favored if both parties offered similar terms. It seemed the whole charade was a well-choreographed show to switch the government's support.

The mayor must have been astonished, as he was blindsided and ambushed by his colleagues. He had presided over the meeting at which the decision had been made to go with Newbridge. Now he voiced his objection to Chinatrust. To the party secretary's argument that whoever paid the highest price should get the deal, he said that the government had already agreed to a deal with Newbridge and Shenzhen should honor its side of the bargain. The mayor also warned that the government would open itself to legal challenges by Newbridge if it breached the agreement.

But the party secretary was the boss. And it appeared that the mayor had been overruled.

After that meeting, Xiao's team once again stopped working with us. Xiao himself was a disciplined and tight-lipped man, and we could not pry from him what was holding him back. We were dumbfounded when we heard from various sources what had transpired during that meeting. It was highly unusual in Chinese politics for the number-one and number-two officials to butt heads like that. And so while we were distressed by the switch of support from the party secretary, we were also surprised that the mayor had taken such a principled position in disagreement with his boss.

Nonetheless, the party secretary had spoken. The decision was made. On May 6, we received a letter from SDB's law firm informing us that the transition control committee we had sent in had been "terminated" by SDB on the grounds that the arrangement had expired. This was basically a formal notice they were canceling our deal. Now that the government had decided to talk with Chinatrust, Zhou did not have to pretend to be working with us anymore. He simply wanted us to go away.

It's hard to describe how I felt at this point. It takes a lot to bring me down, but I felt deeply frustrated now, at a near-total loss in terms of what to do next. My partners were also deeply disappointed, of course.

What options did we have? Not many. We had intimated that we might take legal action against the government. But were we really going to do that? Few thought it realistic that legal pressure would bring the government back to the negotiating table. No one could recall an example of such tactics working in China. That said, we weren't going to give up without a fight.

We still hoped that we could persuade the government to do our deal. The framework agreement required the parties to seek to resolve any disputes through "amicable consultation" for a period of time before resorting to overseas arbitration. If we were to take legal action, we had to follow that protocol.

I had already prepared a letter to the Shenzhen government when we received SDB's formal termination notice. The letter was approved by Newbridge's co-chairmen, Blum and Bonderman, and signed by Dan Carroll and me; it went out to the chief negotiator Xiao. In the letter, we argued that the government had breached the framework agreement in several ways, primarily by talking with Chinatrust. The letter was also intended to show for the record that we had completed the "amicable consultation" phase and were now free to take legal action, if we wished. The letter was addressed to Xiao but intended for Shenzhen's leaders. We hoped the message would persuade the government to reverse course. The letter referred to the framework agreement of June 21, 2002, and said, in part:

> *Regrettably, your side has repeatedly breached multiple sections of the Agreement. We pointed out these problems to your side many times both during formal meetings on February 12, 2003, and thereafter, requesting that we resolve these problems through amicable consultation. However, after months of amicable consultation, the breaches by your side have continued without being remedied. For example, your contacts with Chinatrust of Taiwan regarding the sale of your [SDB] shares constitute a material breach of the Agreement.... In addition, Chinatrust informed us directly that representatives of the Shenzhen Government independently initiated these contacts and discussions.... As you know, we sent a letter to Chinatrust on February 14, 2003, warning that institution to cease tortious interference with our contract and informing Chinatrust that we reserve the right to take legal actions against it.... Therefore, please be forewarned that we have the right to resort to legal actions including initiating arbitration at any time....*

We were, however, still holding out hope:

> *If Shenzhen government intends to discontinue the contact with Chinatrust..., remedy all of the breaches, sincerely honor all of its agreements and commitments with us, and execute the completed definitive documents agreed by both parties, please promptly let us know.*

We delivered the letter in both English and Chinese, and copied the entire leadership of the city government—the party secretary, the mayor, Vice Mayor Chen Yingchun as well as his predecessor, Song Hai (now vice governor of Guangdong Province), Executive Vice Mayor Li Decheng, Deputy Secretary General Yu Weiliang, and Liu Xueqiang, the former deputy secretary general who was now head of a district in Shenzhen.

Although we hoped they would heed our warning, we were not optimistic. It was hard to avoid the conclusion that we had lost the deal. It was a staggering setback to consider, given all our work, and how close we had come to clinching it. We didn't seem to have many options. Almost no one investing in China would want to get into a dispute—much less a legal fight—with the government. But we were unwilling to abandon hope, or to stop trying. We had come a long way. And while our due diligence had confirmed our worst fears about the asset quality of the bank, we were also convinced that we would be able to turn SDB around, given the chance to install our own management and risk management systems. If we walked away now, all our efforts would have been wasted. And our reputation would take a beating.

I knew that many leaders in China's financial industry wanted to see the deal go through. By now it was widely understood that it would constitute a major step, a test case of reform in China's banking sector. Friends encouraged us to persevere. I received a particularly dramatic note from a Chinese friend: "You would let down many Chinese if you don't fight. Now it's time for the bayonet to draw blood!" I understood his metaphor—he was urging us not to give up but to fight. Those in the establishment who supported and encouraged us had not the slightest interest in our commercial success or failure. But they, like the mayor, were reform-minded people and they, too, thought this deal was an important barometer for the country's reforms.

After much internal debate, we made up our minds to fight. It was not an easy decision; we had to consider the possible backlash and the consequences to our franchise in China. To my relief, all my partners, including Bonderman, Blum, and Carroll, were on board with the plan. We decided to wage our battle on several fronts.

The first was to file a lawsuit in Texas against Chinatrust. The second was for Bonderman to send a formal complaint to Chinese prime minister Wen Jiabao—a shorter version of the argument we had made in our missive to Xiao. The letter was delivered through two channels, one via the Chinese embassy in Washington and the other through Sandy Randt, the American ambassador to China. Bonderman's letter read, in part:

> We regret to inform you that Shenzhen has knowingly breached a binding agreement entered into last year and approved by several instrumentalities of the Government of the People's Republic of China.... Despite Newbridge's best effort over many months to resolve any disagreements through amicable consultation and friendly negotiations, Shenzhen has continued to breach a signed agreement, hid the truth from us and frustrated the successful completion of the Transaction.

He concluded this way:

> *Premier Wen, Newbridge is an American firm and our professional conduct is bound by American and Chinese laws. We believe that the Chinese government appreciates the stature of a major international firm like ours and welcomes our participation in the reform of a banking system which is founded on integrity and trust. We will ensure that international best practices be followed after our investment. For many months now, we have hoped that Shenzhen, as a government, would honor its commitments, promises and agreements. But we are sorely disappointed.*

A strong but respectful message, I thought, as it was necessary for us to point out the gravity of the matter, which we hoped would draw his attention.

The third prong of our fight would involve me going to Beijing to complain to regulators and ask them to intervene.

There was a last-resort measure on our minds as well: to bring our complaint to arbitrators outside China, in accordance with our original agreement with Shenzhen. Taking the government to arbitration in a foreign jurisdiction would be tantamount to suing the government. We thought we would do this only if all else failed.

★ ★ ★

It was about this time that an unprecedented epidemic broke out in China. Severe acute respiratory syndrome, better known as SARS, struck Hong Kong in late March 2003. By May the disease, which was highly contagious, had spread to other parts of China. In Hong Kong, not a single day went by without reports of deaths, one after another. As the toll mounted, so did the panic. Schools were shut and streets deserted. The few people who ventured out wore surgical masks. The hustle and bustle of Hong Kong's city life came to a standstill, giving way to an eerie quiet, as people stayed in the relative safety of their homes. Thousands of people fell ill, and hundreds succumbed to the disease. It was a frightening time. It would take months for the disease to be contained.

The epidemic reached its peak in May 2003. No one in the greater China region traveled unless it was absolutely necessary. People everywhere seemed desperate to avoid contact with us. I flew to Beijing on May 7. The flight was almost empty. The few passengers all wore masks. When lunch was served, some passengers refused to eat, too afraid to remove their masks.

I carried with me a big box of files and documents, including all the agreements between Newbridge and Shenzhen, as well as all the submissions SDB and Shenzhen had provided to higher government authorities and agencies. The box was so heavy that I had to lug it all on a foldable luggage cart. The documents would show unequivocally that we had a binding agreement and exclusivity for the deal.

I arrived to find the streets of Beijing deserted. Just as in Hong Kong, the number of deaths and new patients with SARS was reported daily in newspapers and on TV. As the figures were still climbing, social interaction remained almost nonexistent. Beijing had imposed a strict quarantine. Schools were closed, as were some workplaces. Beijing's notorious traffic jams had also disappeared. The traffic was so smooth that I found myself wishing there was some way, other than a major health scare, to keep it that way.

I visited the CBRC, the banking regulator, in the hope of meeting with its chairman. But he was not available. Instead I met with Tang Shuangning, the vice chairman, and Nan Jingming, the director general. I told them that Shenzhen had breached the agreement and made clear that we would sue them if this were not resolved amicably. I was appealing to them to intervene. For all my blunt talk, however, they seemed unmoved. They said the dispute was none of their business; their business was to make sure that whoever bought the bank was qualified to do so. I left the meeting feeling quite dejected. I had made my trek to see them and had achieved nothing.

Other senior people I met in Beijing told me that if the party secretary and mayor of Shenzhen disagreed, it would be difficult for anyone else to be involved or take sides. I impressed upon everyone that we had a strong case against Shenzhen, and I gathered plenty of sympathy wherever I went, but little else. I was told that any hopes depended on whether the big boss, Prime Minister Wen Jiabao, wanted to be involved. Bonderman had already written to him. We had to wait and see.

The Shenzhen side paid no heed to the letter we had sent. Instead, they doubled down by having SDB issue a press release on May 12, announcing the termination of our transition control committee. Now the fight was out in the open. We were confronted with the question of whether to wage a public battle, and, if so, how. I was bombarded by calls from reporters. We issued a terse statement repeating the core arguments: We had a binding agreement, and we fully expected Shenzhen to respect its commitments and honor its contractual obligations.

The press had another field day. Good or bad, it was big news. All the major newspapers covered the breakup of the SDB talks. The *Financial Times* ran the story on its front page. It had been a big surprise when we announced the framework deal in the first place; the market had been abuzz over what it perceived to be a watershed event for China's banking sector. Now the breakup of talks was causing a different stir; the deal's failure was read as a setback for reforms.

We at Newbridge typically avoided the limelight, preferring to keep a low profile and "operate under the radar," as we often said to ourselves. This time around, however, we felt compelled to respond to the press release by SDB and to media inquiries to set the record straight.

The public response from Newbridge must have taken the Shenzhen government by surprise. Even if officials there knew we were in the right, they almost certainly hadn't expected to be publicly rebuffed by a foreign investor. Usually investors would just walk away under such circumstances. Now, to deflect further media attention, the Shenzhen side seemed eager to distance itself from the whole affair. Xiao told the media that the Shenzhen government had been nothing more than a bystander, having had nothing to do with the SDB deal, which he argued was only a commercial transaction between the bank and a foreign investor.

I doubted if any members of the media bought his story. If what he said had been true, then why was he, the president of a different bank, speaking about the SDB transaction? It made no logical sense.

Amidst all this, Xiao asked for a meeting with us.

I was hoping the request signaled a reconciliatory gesture. But here I would be disappointed once more. My colleagues and I arrived at the SDB headquarters just before 3 p.m. on May 13. Xiao wanted a word with me first. He explained that he had been instructed to deliver a message to us in the formal meeting. Xiao was an honest and upright man and he must have felt uncomfortable telling us something he knew to be untrue. It was his way of telling me he had no choice. Then we proceeded to the big conference room where his team and ours were gathered.

After a brief opening, Xiao delivered his official message. He claimed that nobody from Shenzhen had ever spoken with a third party, and that nobody had spoken with Chinatrust. Then he said that the transaction was a commercial decision taken by the entities owning SDB shares, but not by the government. Therefore, this reasoning went, the government itself had nothing to do with it. Everyone in the room, including Xiao himself, knew this statement was false. But we also knew it was pointless to argue.

The government side must have thought that by denying it was the decision maker, it would shield itself from legal challenges. How could we sue an innocent observer that had been on the sidelines? Indeed, all the SDB shares to be transacted were owned by four commercial entities controlled by the government, and technically not by the government itself.

I thought the antics employed by the Shenzhen side were clumsy, almost farcical, but mostly they were just irritating. I also thought they fit perfectly with a Chinese idiom: The thief covers his own ears when stealing bells, in the hope nobody would hear the chime if he himself could not hear it. In this case, others had long since heard the bells chime. In fact, we had a full collection of Shenzhen's fingerprints on the transaction. We were not going to let this official denial of involvement stand.

I took out a stack of documents from my bag. The first was an official submission by the provincial government of Guangdong to the central government. I read from the document a paragraph that stated the city government of Shenzhen had selected Newbridge for the SDB deal. To be fair, we did not know if the government had directed the president of SDB to look for a foreign investor, or whether he had taken the initiative after consulting with the government. It made no difference; the city government controlled the bank and had the power to decide what to do with it.

Xiao appeared stunned, once again, just like previous times when I showed him some documents. He had not known of the existence of this document and other official papers issued by the government prior to his involvement. He certainly had not known we had them in our possession. The documents immediately shattered the government's denials of its pivotal role in the matter. As he examined them all, Xiao fell silent.

Chapter 11

"Nut Case"

O n May 14, 2003, Newbridge filed a lawsuit against Chinatrust in Texas for "tortious interference." In the filing, we asked that Chinatrust "cease and desist" its interference. The lawsuit also named Zhou Lin as "aiding and abetting" the interference. Our petition was soon picked up by the media. Several English-language newspapers, including the *Wall Street Journal*, the *Financial Times*, *BusinessWeek*, and the *South China Morning Post* covered the lawsuit, some on their front pages. Taiwan's *Economic Daily* published the news. A Chinese newsletter, *Major Financial and Economic News Weekly*, also reported the story and specifically mentioned the reference to Zhou.

BusinessWeek ran an article on May 23, under the title *China Banking's Great Wall*, and a subheading that went like this: *Newbridge Capital's plan to run a Chinese bank comes undone—and the U.S. firm is doing the unheard-of by taking the case to court.* The article pointed out the significance of Newbridge's setback:

> *It was the deal that would change Chinese banking forever. When U.S.-based Newbridge Capital Inc. agreed last October to buy a controlling interest in Shenzhen Development Bank Co., Chinese officials were so enthusiastic that they even allowed Newbridge to install its*

own management team before final details had been ironed out. At last, Chinese regulators and bankers seemed ready to allow a real commercial bank to flourish under foreign control.

As for the lawsuit, the reporter wrote:

> *It's rare for even the most frustrated foreign investors in China to turn to the courts, for fear of jeopardizing future relationships. But Newbridge isn't a normal investor. The Chinese "didn't realize they were dealing with a nut case"—that is, a firm that wouldn't play by routine rules, says a knowledgeable source.*

"The knowledgeable source" was yours truly. When we spoke by phone, the reporter was astonished by our moves. "Nobody does business in China," he said, "by suing the government." I had thought that was exactly what the Shenzhen side had calculated; they had expected us to simply go away. I didn't know what to say in response, so I gave him that line. They were dealing with "a nut case," I said. We both laughed.

We weren't suing the government—not yet. But under the framework agreement, either party had the right to file for arbitration outside China, under the rules of the International Chamber of Commerce. The result would be final and binding on both parties, and because China was a member of the New York Convention for arbitration, offshore arbitration awards could be enforced within China, by the country's highest court. However, this would be our last resort.

★ ★ ★

I did have other Newbridge business to attend to, though it seemed sometimes that the SDB deal—or nondeal—had become all-consuming. As we filed the lawsuit against Chinatrust, I was preparing for a trip to New York, for an arbitration hearing, this one involving Korea First Bank (KFB). Newbridge had acquired control of KFB from the Korean government in 2000. That agreement required arbitration in the event of disagreement over the amount that KFB asked the government to pay for certain impaired legacy loans. Such hearings were predictable, because the Korean government felt politically compelled to dispute any amount of payment calculated by KFB, but in the first two years these had been set-tled in our favor right before the hearing. This annual ritual notwithstand-ing, the relationship between the two sides was friendly and professional. The Korean government had honored its obligations under the contract.

I arrived in New York on May 23 and checked into the Omni Berkshire Place, in the heart of Manhattan and just a few blocks from Central Park. I thought my colleague Daniel Poon and I were going to meet with our lawyers the next day to prepare our testimonies for the KFB arbitration. But as soon as we arrived, we were informed that nobody wanted to see us, or anyone from Hong Kong, due to the fear that we might be carrying the contagious SARS virus.

Although we could not have business meetings, we could move freely away from our hotel and eat in any of New York's restaurants. I con-sciously avoided seeing anyone I knew in case they might have the same concerns as our lawyers. New York was unusually chilly that week, at least for the late spring. It was about 50 degrees Fahrenheit, and it rained nearly every day. Between the weather and the SARS paranoia, I spent a good deal of time in the hotel room, reading or working on my laptop.

Even in self-imposed semi-quarantine, it was nice to be in New York City, and it felt like a welcome break from the nonstop, nerve-wracking SDB negotiations. The time difference between New York and Hong Kong was 12 hours; Hong Kong woke up just as the sun went down in New York. To synchronize my days with all the activities back in Asia, I did my best to keep Hong Kong time, since I couldn't see anyone in New York anyway. I worked, often on conference calls, through the night, going to bed at sunrise. My only breaks came in the afternoon when I went to Central Park to run or to ride horses.

I had learned to ride during my time in the Gobi Desert and then in Hong Kong's Jockey Club, and still enjoyed it whenever I could. There was a stable in New York City, the Claremont Riding Academy, on Manhattan's Upper West Side. It kept some fine horses in a multi-level stable in which the floors were connected with a winding ramp designed for the horses. I had never seen horses stabled this way. It reminded me of how cars in New York are frequently parked on top of one another in elevated racks. You could rent a horse there and ride it for a couple of blocks through busy Manhattan traffic into Central Park. My horse, named Ronin, doubled as a show horse in the opera because of his magnificent build. It felt great riding around the bridle path that circled a beautiful reservoir in the park.

But even pleasant rides couldn't take my mind off SDB. While I was in New York, I learned that Mayor Yu had been transferred away from Shenzhen to become the vice governor of Hunan Province. We had counted him as a strong supporter of our SDB deal. With him out of the picture, the prospect of resurrecting our deal looked even bleaker. I also learned that the banking regulator had asked Zhou Lin to go to Beijing to explain what was going on with the SDB sale, but that the Shenzhen government had told him not to go. It seemed that the Shenzhen side was digging in its heels.

Our options were limited. Even if we could stop Chinatrust or any third party from stepping into our shoes, we were unlikely to be able to force the government of Shenzhen to sell SDB to us. I was growing more pessimistic, as were my partners. I met Bonderman for breakfast in New York on Friday, May 30. He thought the SDB deal was dead. I had to agree with him. I could see that both Blum and Bonderman were now doubtful that the lawsuit against Chinatrust would achieve anything. They were also wondering where it might lead us. We certainly did not want to make Chinatrust our nemesis, if neither we nor they could achieve our objectives anyhow.

However, we were not prepared to throw in the towel, either. I thought, as with anything in life, "man proposes, God disposes;" and so we had to keep "proposing." If we gave up, I thought, God would surely forsake us. Therefore, we had to make every effort. If eventually we failed, we would have no regrets, knowing that we had tried our best.

We knew that many influential people in Shenzhen and Beijing, including key regulators, still wanted to see a deal of this nature happen. We also knew that the media, both international and Chinese, were generally sympathetic to our cause. That counted for something. It was actually quite remarkable that although we were fighting with a municipal government in China, the Chinese press had generally applauded the deal, or at least reported factually about the fight. No Chinese newspaper had printed anything that blamed Newbridge for the impasse. The international press still showed a keen interest.

On June 10, the *Wall Street Journal* ran an article with the headline "Newbridge, SDB Deal Went Awry, Shedding Light on China's

Reform." I thought the reporter provided a good summary of where the matter stood:

> In mid-May, SDB renounced the contract authorizing Newbridge to manage the bank for reasons that haven't been fully disclosed. A few days later, in a Texas district court, Newbridge sued Chinatrust Commercial Bank, controlled by members of the prominent Koo family of Taiwan, alleging that Chinatrust had interfered with their deal by holding talks with Shenzhen Development Bank despite being informed that Newbridge had an exclusive right to talk to SDB, according to the suit.

I thought the insight in the piece was that the key to breaking the logjam rested with Shenzhen, and that we at Newbridge were in a balancing act—between pressuring the city government with legal action and hoping to find a face-saving way to bring it back to the table:

> Chinese authorities have indicated their support for Newbridge; the country's top regulator, Chinese Banking Regulatory Commission chairman Liu Mingkang, in May affirmed the central government's approval of the deal. But some observers say the travails of Newbridge with SDB show the pitfalls of top-down deals, as Beijing tries to impose its preferred outcome on local fiefdoms.

> If the deal with SDB falls through, Newbridge may bring legal actions against both the Shenzhen government and SDB for breach of what it claims is a legally binding agreement, according to the complaint. But Newbridge seems eager to avoid a direct confrontation with the Chinese side in the hope of finding a way to complete the transaction, according to people familiar with the matter.

★ ★ ★

We were still awaiting news on two fronts: our lawsuit against Chinatrust, and any response to Bonderman's letter from China's prime minister. For the moment, we were reluctant to escalate the fight by taking the city government to arbitration. And in July, we saw encouraging signs that the top level of the Chinese government was at least paying attention.

On July 9, I met with the CBRC chairman Liu Mingkang in his office in Beijing. I had sent a message that I was not going to talk about SDB; I was afraid he might refuse the meeting if SDB was on the agenda. To my delight, he brought up the subject almost as soon as we sat down. He said that he had wanted to talk with me earlier but heard I had gone to America.

Liu told me that the prime minister had indeed received Bonderman's letter via the Chinese ambassador to the United States. The PM had delegated Huang Ju, the vice premier whose portfolio included banking and finance, to investigate. Huang Ju had asked that the CBRC to review Bonderman's letter, look into the matter, and report back to the vice premier. That was the reason Liu had wanted to see me. That alone was promising.

Both of the top banking regulators, the central bank and the CBRC, had indicated their support for the transaction. However, Liu Mingkang said that the official position the PM and vice premier were advised to take would be that the central government and regulators were in favor of the transaction, but that it was a commercial matter that needed to be sorted out between the transacting parties. The CBRC had separately communicated its support for the transaction to the party secretary and the mayor of Shenzhen.

The CBRC had also conveyed its views to the city leaders that SDB was poorly run, and that the banking regulator would examine its operations. And then there was this important note: The CBRC made it clear to Shenzhen that it would be "very difficult" for it to approve a Taiwan bank—a clear reference to Chinatrust—to take over SDB. There was no official relationship between the regulators of the mainland and Taiwan, so they had no official mechanism to coordinate on regulatory matters. I knew that when a Chinese regulator said *very difficult*, it really meant *impossible*.

This was unvarnished good news for us. It meant that Chinatrust was effectively out of the picture as a legitimate competitor. It did not necessarily mean, however, that Shenzhen would invite us back.

★ ★ ★

Our resolve to fight the city government was wavering. It seemed that Shenzhen had moved on. The city officials reshuffled the management of

SDB, promoting Zhou Lin to chairman and appointing another man as president. Zhou had been running the show as president, and he was still doing so as chairman. There was no sign that Shenzhen wanted to return to the negotiating table.

A number of people close to inside sources told us that the deal was hopeless, "dead" even, which was consistent with how we felt as well. But other people encouraged us not to give up. There was nothing to lose, it seemed, and so we decided to escalate. It was time to turn to our last resort.

On September 16, Newbridge's lawyers filed for arbitration against the Shenzhen government with the International Court of Arbitration in Paris. I promptly informed Xiao, the chief negotiator, of the action we were taking.

The filing kicked off a flurry of activities on Shenzhen side. A number of third parties were asked by Zhou Lin and others to mediate or to put pressure on us, and on me specifically, to withdraw the arbitration. But no mediators could offer any solution to break the impasse. I heard later from Xiao that the government officials had been very upset (no surprise there) and wanted the arbitration request withdrawn. They had taken a stand: Shenzhen would talk with Newbridge only if we withdrew the arbitration first. We countered that we would withdraw the arbitration request only if Shenzhen came back to the negotiating table. Neither side would yield. And so the stalemate continued.

Some people close to Dick Blum reached out to him, trying to persuade him to accept some kind of a settlement, but without specifics. The purpose obviously was to pressure Newbridge to back off. Blum had doubts as to whether legal action would get us anywhere, but he did not want us to give up without a fight. The pressure from these back channels only stiffened his resolve; now he was firmly of the view that we should settle for nothing short of completing the proposed transaction.

There was one move that hit close to home. An intermediary passed a message to Blum from the city government suggesting Newbridge remove me from the deal, in exchange for the Shenzhen side resuming negotiations. Blum flatly refused and stood firmly by me. Blum is a principled man, and he would not sacrifice his partner to save a deal. No chance, he said.

I knew that some on the Shenzhen side were upset with me; they thought, correctly, that I was chiefly responsible for creating all this trouble for them, the arbitration filing included. But our intention had not been to embarrass the government and we had tried everything possible to avert it. All we wanted was to get back on track with the deal. I signaled that I would be happy to step aside if Shenzhen was willing to resume the SDB transaction with us. I personally did not mind staying in the background if that would help save the deal.

On November 6, Au Ngai and Ricky Lau, along with a former partner of Newbridge playing the role of a mediator, went to Shenzhen to meet with the party secretary and the new mayor, Li Hongzhong. As we all knew that I had become persona non grata, I stayed behind.

I met with the trio over dinner in the Conrad Hotel after they returned to Hong Kong that evening. They reported that the party secretary had complained bitterly about me for almost two hours. It was clear that her complaint was based on misinformation. I didn't mind being the punching bag if I could help ease the way for Shenzhen to return to the negotiating table.

But it was not so simple. Shenzhen continued to insist that we withdraw the arbitration first, before any resumption of talks. We saw no reason to do so; arbitration was perhaps the only leverage we had.

Then in November, Shenzhen softened their stance a little. In the middle of a lunch, my colleague, Au Ngai, received a call from Yu Weiliang, the deputy secretary general. Yu offered something of a compromise. He wanted a letter from us saying that we would *consider* withdrawing the arbitration request. I thought that was okay, as long as Shenzhen agreed in return to resume negotiations with us. After much back and forth, the Shenzhen side indicated its willingness to meet with my co-managing partner, Dan Carroll.

We thought this signaled a welcome thaw in the government's position, and that Carroll could at least use the meeting to gauge how serious the officials really were. But just as Carroll was about to leave for Shenzhen, word came that they would *not* meet with him unless we withdrew the arbitration. It seemed that those who still wanted to play hardball had prevailed over those who were inclined to be more conciliatory, but it seemed to me that their ranks were cracking.

The Shenzhen side simply wanted to will the arbitration away. When we refused to budge, they thought the arbitration could be ignored. When they finally realized it could not, they engaged lawyers to handle it. But Shenzhen's lawyers had no experience in international arbitration. The arbitration court imposed a deadline for Shenzhen to respond by November 22. Its lawyers sent a request to the court just days before the deadline to ask for an extension to December 22. The court rejected the request and gave Shenzhen until November 28 to comply. Then Shenzhen simply fired their lawyers, presumably because the lawyers had failed to stop the proceedings. The government then hired another law firm, which fared no better. Our contract with Shenzhen required English language for arbitration. Either because the Shenzhen lawyers could not function in English or because they wanted to make it difficult for the arbitrators—we never knew which was the case—they submitted their answers in Chinese, no doubt much to the chagrin of the arbitrators who did not read Chinese. We watched their clumsy moves with amusement, as they seemed, unintentionally, to be doing everything possible to help our case.

Chapter 12

Untying the Knot

As of 2004, it had been nearly two years since the whole process had begun, and we were still far from breaking the deadlock. Nevertheless, soon after the new year I met with a regulator who told me there was still a strong desire at the top levels of the central government for our deal to proceed. He suggested that one of our co-chairmen write a letter to China's leaders to express our commitment to the Chinese market and our intention to bring international best practices to SDB. The letter, the official said, should include our credentials and a description of our first-rate team, its experience in banking and so forth. He also suggested we mention Shenzhen's wish for us to withdraw the arbitration request, our understanding of the same, and our own desire to work out a deal in an amicable manner.

I took the message to indicate that the Shenzhen side was ready to resume talks but didn't want to appear to have capitulated to Newbridge. They wanted a face-saving way to get the deal back on track. I understood this, particularly given that our quarrel had been so public. We were willing to oblige. After all, we just wanted to do a deal, not to win a fight, and we were eager to get back into the good graces of the Shenzhen government.

★ ★ ★

Few deals proceed without detours, and ours certainly had its share. Just as we were preparing to return to the negotiating table, a Hong Kong company contacted us, presumably also at Shenzhen's request. We knew the firm well. The chairman was suggesting he could bring us back into the deal through a joint venture between Newbridge and his company.

The proposal was for the two parties to form a special-purpose vehicle (SPV), of which each would own 50%. Then the SPV and Newbridge itself would own 60% and 40%, respectively, of the SDB stake in question. That way, on the surface of it, Newbridge would only be a minority shareholder of the SDB stake, owning only 40%, but indirectly we would own 70%, as we would also own 50% of the SPV. The chairman promised us that Shenzhen would be happy with this structure and willing to move forward.

I knew that the Shenzhen side still couldn't get over the face issue. I was willing to be the fall guy, taking the blame for all the conflict and staying in the background. But apparently that wasn't enough. Someone else had to appear to take Newbridge's place.

I thought the joint venture was a clever idea but perhaps too clever by half. I doubted that the Shenzhen side or the chairman of the Hong Kong company had considered whether the banking regulator would approve any nonfinancial firm taking ownership, let alone perceived control, of a Chinese bank. I was quite certain, based on our own experiences, that the regulator would balk at this.

On February 6, I wrote a memo to the chairman of the Hong Kong company, listing all the terms we had previously negotiated and agreed to with the Shenzhen side. "If a deal can be structured to include all these terms," I wrote, "Newbridge is prepared to close the transaction...." In other words, their name—but our deal. And face would be saved.

Although I remained in the background in deference to Shenzhen's wishes, chief negotiator Xiao and I kept a back channel of communication. By late February, Xiao was telling me that Shenzhen was ready, again, to proceed with the SDB deal. Xiao was making an effort to dig both parties out of the hole we were in and move things forward.

Now that Shenzhen was willing to resume negotiations, we decided to withdraw the arbitration request. When I informed the CBRC chairman, he was delighted. I followed up with a formal letter to the CBRC on March 25, 2004. The latest round of skirmishes was over. Now we would see whether our move would be reciprocated.

Here we were, seemingly on the brink of significant progress. Now the Hong Kong company said it was no longer content to be a minority shareholder. On March 31, its chairman informed us that he wanted 50% actual ownership, instead of the 30% in our initial agreement. At this point, Zhou Lin must have seen the writing on the wall and understood there was an official desire—either in Shenzhen or in Beijing, or perhaps both—to bring the deal to a conclusion. He came back to our side, supportive of the transaction once more. But he also wanted the Hong Kong company to own a majority of the shares of the SPV. We could not agree. Although the two parties would be friendly partners in the deal, control was critical to allow us to lead a successful turnaround of the troubled bank.

Then, almost as suddenly as they had joined the fray, the Hong Kong company abruptly informed us it was dropping out entirely. We assumed they either didn't want to do a deal without control or thought we wouldn't be able to conclude the deal without them. Maybe they were calling our bluff, and simply hoping we would agree to their conditions.

But at this point, for reasons we were not entirely clear on, the Shenzhen side seemed strongly motivated again to get the deal done with Newbridge and was no longer concerned about a façade of someone else fronting the transaction. Now that the proverbial train was leaving the station, with both parties on board, Shenzhen's priority was getting the deal done without further delay or complications.

As I continued to negotiate details with Xiao behind the scenes, my colleagues Au Ngai and Ricky Lau traveled almost every day to Shenzhen to work on the documentation with Xiao's team. The major terms had already been agreed on before Shenzhen walked away. Neither side wanted to reopen major issues. The price would still be 1.65 times stated NAV. The most recent NAV reported by SDB was about 4 billion yuan or about $500 million, so 1.65 times of it would come to $830 million. Shenzhen would sell about 18% of the total shares to Newbridge, which would be worth about $150 million.

Xiao must have received marching orders from the city government. He was now acting with urgency, although as was too often the case, this positive development came with trade-offs. So, when on April 1 he made a request that Newbridge be prepared to sign the final documents, he also said we should be ready to forgo any further due diligence. Continuing with our diligence process, he said, would only create fresh uncertainty; and he insisted that the Shenzhen side wanted only to conclude the deal without further delay.

This was good and bad news, as it created yet another conundrum for us. We certainly did not want to risk losing the deal again. But we didn't want to step into a minefield, either. Our transition control committee had been kicked out of SDB about a year ago; we had no idea what might have happened to the loan book and assets of the bank since then. We had good reason to worry that more bad loans had been made (we would later be proved correct). But we had no clue how big the problem might be. We would have to take a great leap of faith to proceed with the transaction without further due diligence.

★ ★ ★

Internally we debated whether to accept Xiao's demand. When Au Ngai, Dan Carroll, and I had a conference call to discuss the issue, Au and I were in agreement that time was of the essence; we would risk the deal's collapse if we did not seize the moment, and the momentum. We advocated for taking the risk and taking the deal without further due diligence. If you were infested with lice, I reasoned, more or less of them would not make much of a difference. The same was true with bad loans. So, we should proceed to close the transaction. Carroll felt further due diligence was needed, just to be sure we weren't sinking into a hole from which we could not recover. We could not come to a consensus.

In such instances it was necessary, and always helpful, to track down Bonderman. I spoke with him that night by phone and gave him my take on the issue and the situation. It didn't take long for him to say that he was with me on this; he would be agreeable to forgoing further due diligence, if we believed there was no other way to get what we wanted.

Meanwhile, Xiao and Zhou Lin were working together to get the green light from the CBRC to allow Shenzhen to proceed with the signing of the final documents. The CBRC made only one request, which was that Newbridge agree to be locked in the investment for at least five years. I had verbally indicated as much to Shenzhen a year before, so it was not an issue for us. We had only one condition of our own, and it was critical: We had to be able to control the bank.

I arranged a call with the CBRC chairman on May 12. He had no issue with our proposed language that "under the pre-condition that Newbridge can exercise all the rights as the controlling shareholder, Newbridge shall not sell or transfer the shares of SDB … for five years." Further, he confirmed that the CBRC would permit foreign nationals to assume the positions of chairman and president of SDB, as we expected to make such appointments. This was also vital for us, as we wanted to bring the best talent into those positions regardless of nationality. In this regard, our interests and those of the regulator were aligned.

Now, once again, our hopes were high.

On the next day, May 13, the PBOC made an announcement imposing a differential deposit reserve ratio on SDB, requiring SDB to maintain a higher ratio with the central bank than its peers. This is a tool of a central bank's monetary policy. All commercial banks are required to keep a percentage of their deposits with the central bank, which is the reason the central bank is often referred to as the bank of banks. If the deposit reserve ratio increases, banks would have less available funds to make loans. So, if the central bank wants to tighten money supply, reducing the amount of money in the system, it can increase the deposit reserve ratio. If it wants to expand money supply, it can lower the ratio. Usually, there is one required deposit reserve ratio for all banks or for a class of banks, say higher for large banks than for small ones. But it is extremely rare for the central bank to impose a different reserve ratio on a specific bank, which would reduce that bank's ability to lend.

Why single out SDB? The PBOC described SDB as the worst offender in the first quarter of 2004 for its "too rapid loan growth"; its loan book had grown more than 60% in the period. A 60% increase in one quarter! That felt like giving money away. None of us knew how many of those loans had been made to borrowers with weak credit. One thing we did know: SDB had run out of liquidity in 2003 due to rapid loan growth and had been forced to ask the PBOC for funding support—borrowing from it. Naturally, all these problems made us worry that if we did not get into the bank soon enough, it might be too late to save it. At some point an infection really could get serious enough to be fatal.

The signing of final documents was scheduled for 10:30 a.m. on Saturday, May 29, in Shenzhen. I had not spoken with Zhou Lin since the talks had broken off more than a year ago, but I called him that morning. He sounded pleased to hear from me. As it happened, he was just about to chair an SDB board meeting to approve the Newbridge transaction; he said he was confident of board approval. Au and Lau traveled to Shenzhen to sign the final papers. I stayed in Hong Kong—my presence still sensitive—anxiously waiting for news, hoping nothing would change at the last minute. It was not until about 12:30 p.m. that I saw an email from Au informing me that all the documents had been executed. Zhou also called to tell me everything had gone as planned.

It was a great moment, and a real milestone for us. The day had drawn the curtain on our negotiations with Shenzhen. It had taken more than two years to get to this stage, since I had first met with Zhou Lin.

But it was not over yet. Sometimes I wondered if it ever would be.

★ ★ ★

The deal still needed to be approved by the regulators in Beijing, including the CBRC, even though it had given its consent for the parties to execute the final agreements. There was also a lingering uncertainty about whether the deal would ultimately receive all the necessary approvals. But I hoped the toughest part was behind us.

SDB's announcement of the signing, which I had helped draft, hit the press on Monday, May 31, 2004. The *Wall Street Journal* was the first one to cover the story that morning, based on a tip we had provided. The news took the market by surprise and caused quite a stir in both Chinese and foreign media; the market had long considered the Newbridge–SDB deal dead. It had been a big surprise when the deal was announced, two years earlier, and another surprise when the deal collapsed, a year after that. And now there was a new twist. I hoped the third time was the charm and we would not surprise the market, or ourselves, again.

We generally declined to speak with reporters. We were sensitive to the Shenzhen side, given the history of the negotiations, and did not want to appear to be taking a victory lap. And, as we surely knew by now, it was not over until it was over. It was still too early to celebrate; we still needed all those approvals. Nonetheless, the international press was enthusiastic. "The first foreign takeover of a Chinese bank will be hard to copy," observed *The Economist*:

> *After a two-year courtship that veered from friendly negotiations to broken promises and an ugly international lawsuit, Newbridge Capital, a feisty American private-equity firm, has beaten the world's biggest financial institutions to become the first foreigner to gain control of a Chinese bank. HSBC, Citigroup, Standard Chartered, the Asian Development Bank and the IFC, the private-sector arm of the World Bank, have also been scrambling to buy stakes in mainland lenders. But so far only Newbridge has managed it.*

We couldn't have written the story better ourselves. And yet I couldn't help but notice a shadow of foreboding in the article's final line: *"The first deal to take control of a Chinese bank may also be the last for quite a while."*

★ ★ ★

Two months after we submitted all the documents to the PBOC and the CBRC for approval, we were still waiting to hear from them. We wondered about the delay, and worried that SDB continued to accumulate risks as it conducted its business as usual. After we had signed the final documents, we sent the members of our former transition control committee back into the bank, but we did away with the group's formal name. Like before, our team only had the power to veto sizeable loans that came to the attention of the credit committee at the headquarters level; it had no power or any knowledge of the loans made at branches or of other risks that the bank might be taking. All the clues suggested that "business as usual" for SDB was, to put it mildly, a problem. We had no idea how long it would take for the regulators to complete their reviews and grant us the approvals, although we remained reasonably confident that they would.

Coincidentally, the CBRC official in charge of our case and I had an acquaintance in common: Wang Ji, the first chief negotiator Shenzhen had appointed. Both men hailed from Tianjin, a major city near Beijing. The official knew Wang had dealt with me, and he asked Wang to privately check out our intentions with SDB. The official was unsure whether we could be trusted to control and manage the bank. Even though Wang and

I had failed to reach agreement in our negotiations two years before, there was no loss of mutual respect. So, when he approached me to ask questions on behalf of his CBRC friend, I welcomed his help in getting our message across. The regulator also seemed to trust his judgment.

Wang had gone through some rough patches since we'd last crossed paths. While he was president of Shenzhen Commercial Bank, he had clashed with the firm's chairman, and had been sacked. That hadn't been the worst of it. Soon after, Wang was placed under investigation for corruption. After two years of exhaustive investigations, however, he had been exonerated. This was rare; it was almost unheard of for someone accused of graft to be found clean. Now Wang's name was cleared, but his career had been ruined and he was out of a job.

I wrote Wang an email on July 31 in response to a number of inquiries he had passed on from the regulator. The email was addressed to him but the intended audience was his friend. Although we both knew he would share my message, by writing to him I felt I could be more blunt, dispensing with niceties. I also wanted to alarm the CBRC into action, to get them to approve the deal expediently:

> First, I hope the approval by the CBRC will not take too long.... If the approval takes a very long time, at a certain point, the risks for this bank will exceed the level any investors are able to bear. We know nothing about what has been going on within this bank for over a year now. During this year, its new loans have substantially increased, leading to severe capital shortage and liquidity crunch, to such an extent that it had to borrow from the PBOC. Even under the double supervision of the CBRC and the PBOC, it took a "great leap forward" this year, growing new loans more than any other commercial bank in the country. This has led to the PBOC imposing a differential deposit reserve ratio on it.... Such risks cannot be controlled just by the CBRC talking to and warning them.

> Second, in my view, the bank cannot be in any worse shape. It is broken. In any other country, it would be declared bankrupt and taken over by the government. In view of the size of its bad loans, its reported capital is already a joke which only fools others as well as themselves. It is the liquidity provided by the PBOC that has allowed it to continue to

operate. In view of such, any investor without experiences in managing banks may not have the guts to own it even if given to the investor for free, let alone for 1.2 billion yuan. We dare to own it and to pay this price because we have confidence … based on our experiences.

My message was well received by the CBRC official. "Reasonable and correct" was the feedback from Wang Ji. I was encouraged and hoped the approvals would be forthcoming soon.

But once again, it turned out I was too optimistic. Apparently, the CBRC considered this case to be too consequential to be approved without the blessing of the State Council. Getting the cabinet's approval would increase both the uncertainty and the waiting time. By the end of August, the CBRC was still asking questions about who Newbridge's regulator was in America, why Newbridge did not have consolidated financial statements of all its portfolio companies, and where our entity holding the SDB shares would be registered. I explained that Newbridge was not regulated because it was not a bank; we did not have a consolidated financial statement because the limited partners of Newbridge for each of the companies we had invested in might be different; and finally, that Newbridge was a U.S.-registered company.

Would the State Council give us final approval? I had heard that there was already a meeting of the minds among the governor of the PBOC, the chairman of the CBRC, and Huang Ju, the vice premier in charge of banking and finance. But this would require the sign-off from the prime minister himself. And there was still no telling how long it would take for all this to happen. Or whether final approval would ever be granted.

★ ★ ★

That spring we had sent a team to SDB to assume credit control, just as we had done with the transition control committee team two years before. We had every intention to replace the old management with a new team of our own choosing, and to appoint our own candidates to the board of directors and the supervisory board to take control of the bank. We had been busy identifying candidates for those positions. By the time the approval process was nearing its end, we were more than ready.

The team we had sent into SDB soon discovered one highly questionable loan, a loan that had been extended only recently. The loan in question was for 1.5 billion yuan (~$180 million), and it had become a bad loan almost as soon as it went out the door. To put this in perspective, the reported net capital of SDB at the end of 2004 was only slightly more than 4 billion yuan. This one bad loan alone represented more than 35% of the bank's capital! In any other market in the world, this one loan could have brought down the entire bank.

The loan had been extended to a newly formed company whose stated mission was to provide internet access to Chinese farmers, but the borrower had no assets, no capital, and had never even started the business in question. Incredibly, ownership of the company had been transferred and its mission changed to real estate development. By then, most of the money was already missing. This loan smelled of corruption through and through. We code-named it "B&B" which stood for "Big and Bad," and immediately began an effort to collect on it—even though we didn't yet own the bank.

But it soon became clear that chances for its recovery were slim. In fact, after numerous efforts including lawsuits that would ultimately last more than 10 years, long after Newbridge had exited from our investment, the bank was able to recover the loan in June 2015. Under our ownership, I personally led the effort to recover this loan, but to no avail. There was no doubt in my mind that the loan had made some people very rich, but we never found out who they were, and to our total frustration, everyone we dealt with complained of being a victim, because they got much less than what they had bargained for by having taken over the borrowing entity.

Was this B&B an isolated case for SDB? Hardly. In this episode of our due diligence, we would unearth many other bad loans made by the bank over the years, although none quite as blatant, outrageous, and enormous as this one. If anything, B&B was emblematic of the state of affairs of China's state-owned banks and how they were managed. Basically, the government was an absentee owner, and the management controlled the bank without having to answer to anyone. The quality of a bank was left mostly to the conscience of its managers, and only secondarily to their competence or lack thereof. Control was so weak that some of the money was simply siphoned away by unscrupulous employees or by phony borrowers who colluded with bank employees. For sure, there were good bank managers, but there was no lack of bad apples.

The worst case of theft by bank insiders in China was uncovered in 2001. In that particular case, the perpetrators were managers of a small sub-branch of Bank of China. They made off with an astounding $483 million.

This incident became known as the BOC Kaiping Case. Kaiping is a small town in Guangdong Province. When the Bank of China (not to be confused with the central bank that is the People's Bank of China) finally connected its computer systems nationwide in early 2001, staff at head-quarters immediately saw a massive deficit that could not be reconciled. It took some effort, including an investigation by the police, to discover that three successive managers of the sub-branch had conspired to spirit a total of $483 million into their own offshore accounts. The fraud had been going on for years. When the managers realized they were about to be caught, they fled the country to join their families, who had long since moved overseas in anticipation of this day of reckoning.

It would take years, and a coordinated effort by law enforcement in China and the United States, to bring the lead culprit back to China to face justice. He received a 12-year jail term, the maximum sentence, which was a condition for the United States to agree to extradite him—the Chinese penal code would have been much more severe, up to death, if he had been caught in China. The former president of BOCHK, Liu Jinbao, was sentenced to death with a two-year reprieve (which was eventually commuted to life imprisonment) for embezzling less than $3 million.

Banking reforms would be absolutely necessary to stop such abuses. China's leaders had made the wise decision to carry these policies through. In retrospect, as hard as it was to try turning around a bad bank, it was nothing compared to attempting to turn around a bad banking system. China was lucky to have such men as Zhou Xiaochuan, the central banker, and Liu Mingkang, chairman of the CBRC, in the right place at the right time to take on these reforms. They were supportive of the transformation measures we took at SDB, so we were lucky to be taking over SDB at the right time.

All that said, the discovery of B&B just two months before the closing had given us pause and caused much handwringing at Newbridge. We wanted the government to indemnify us against this bad loan. We also talked with the Shenzhen side to win its support and help obtain approval from the China Securities Regulatory Commission for the bank to raise capital via the contemplated rights issue to replenish the bank's capital before we closed the transaction. In the end, we decided to proceed with the closing once approval from the State Council was handed down, without waiting for the rights issue, because there was no telling when the CSRC would approve it. Any delay would only further increase the risks associated with SDB. The best thing for us to do was to get in there and take control as soon as possible.

Xiao was fully aware of all the problems, including the B&B. He was receptive to my proposal for the government to help mitigate New-bridge's risks. On November 29, 2004, I met with him in his office. After some discussion, he and I agreed on three points. First, Shenzhen would provide Newbridge with an indemnity, or a guarantee on bad loans like the fearsome B&B. Second, part of the proceeds received by Shenzhen from the sale of SDB shares would be loaned to us to subscribe to new shares to be issued by SDB. This ensured that Newbridge would not be diluted by the rights issue. And third, the remainder of the proceeds received by Shenzhen would be used to buy a convertible bond issued by SDB with a 30-year tenor, with conversion rights given to Newbridge.

A convertible bond allows the holder to convert the bond into the shares of the issuing company at a specified strike price. If the bondholder chooses not to convert the bond into shares, he can get the money back upon the bond's maturity. Typically, the conversion right is attached to the bond. But in this case, we proposed that the government would keep the bond but would give the conversion right to Newbridge. In effect, this would give us an option to buy more shares of SDB at the strike price.

While Xiao lent a sympathetic ear to my ideas, he reminded me that he still needed to get the Shenzhen government to agree. The tone of his voice, however, suggested that he wasn't hopeful and, in the end, we received none of these supports.

Meanwhile, we turned to the matter of risk control. At a meeting of Newbridge's partners that November, we instructed our new transition management team to adopt an interim credit policy. Our team had the right to reject any new loans. Our interim policy for the bank was to roll over performing loans when due with a payment plan; we would refrain from extending new loans until we closed the acquisition, which we thought would be the best way to stop the bleeding. Meanwhile, our own people would immediately join the SDB board and assume risk control, even before the closing of the transaction.

All through this period, our competition remained hot on our heels. On December 3, we learned that He Ru, the new president of SDB, had brought Ping An Insurance Group to the city government as a potential buyer. Xiao was worried but also used this as leverage to push us to move quickly to close the transaction without waiting to receive confirmation of the various measures of protection we had asked from the government. He warned that if Newbridge did not proceed immediately to closing after receiving the final approvals, it might lose the deal to someone else.

I found myself thinking of Qian Zhongshu, author of the famous novel *Fortress Besieged*. Written in 1947, it's a satirical story that likens marriage to a fortress. Those inside the walls see all the problems and want to get out, whereas those outside the walls see only romance, and aspire to get in. In a way, we at Newbridge were inside the fortress, struggling with new issues we uncovered almost daily. Meanwhile, many in the market thought we had captured a great prize, and were making every effort to squeeze in and kick Newbridge out. We were so torn—troubled by the problems that were so evident at the bank, yet still contending with competition for the deal, even at this late stage.

Part III

Transformation

Chapter 13

Making History

W e finally received all the approvals, not a moment too soon. The closing took place on December 30, 2004. The entire process had taken 30 months, from start to finish—double the time we had spent on the Korea First Bank transaction. SDB hadn't required nearly as much in the way of actual negotiation as the KFB deal had, but I felt that the strain had been much greater, with all the ups and downs, the waiting games, and uncertainties and anxiety. It gave all of us at Newbridge great relief and comfort that the deal was finally done. And we really had made history. Given the grind of all the work, day in and day out, some of us might occasionally have lost sight of one basic fact: It had never happened before. No outsider had ever taken the reins of a Chinese national bank.

But if my Newbridge colleagues and I allowed ourselves a dose of pride, and some quiet celebration, the closing itself passed without fanfare. It was almost anticlimactic. The work had been done. There was no ceremony; the closing was processed quietly by the finance staff of Newbridge. On the day itself, I was in Malaysia chairing the last board meeting of Newbridge's holding company for KFB. We closed the year having agreed to sell our control in KFB to Standard Chartered Bank and having bought control of SDB from the Shenzhen government. Both of these events took place in the last week of 2004.

Why no big celebration? Why no champagne? The reasons were simple. First, for months we had believed the moment would come; even the press by now had reported frequently enough on the possibility of a deal so when the moment came, it was not the blockbuster story it might once have been. Second, as difficult as it had been to gain control of SDB, we expected an even bumpier road ahead. I recalled the moment when we had finally signed the final agreements for KFB, after 15 months of a long and arduous negotiation, and how relieved I had been when I called David Bonderman to inform him of the good news.

"Congratulations," he said. "Now the hard part begins."

I laughed and responded with a metaphor: "Now that we have slaughtered it, we will have to eat it."

It had proven difficult, but the hard work of turning around KFB had eventually borne great fruit. I was looking forward to the same experience with SDB, despite our understanding that SDB was a far messier problem to fix.

As by far the largest shareholder of SDB, Newbridge would control its board of directors and install a new management team, replacing the chairman, president, and others. Other than Newbridge executives, we identified a number of highly accomplished professionals to serve as directors and supervisors. Publicly listed companies in China have both a board of supervisors and a board of directors, similar to the system in some European countries. The supervisor's role is to make sure that the board of directors discharges its duties in accordance with established rules and regulations. The board of supervisors was usually less powerful than the board of directors, but its voice still carried weight in the decision-making process.

The most important position would be the bank president. After interviewing a number of candidates, we decided to offer the position to Jeffrey Williams, an American who spoke fluent Chinese. Tall, mild-mannered, and soft-spoken, Williams came with impeccable credentials—a bachelor's and master's degrees from Harvard University and years of experience working for Citibank, American Express, and Standard Chartered Bank, mostly in Taiwan, but on the mainland as well. We needed someone familiar with international best practices and, at the time, there were few Chinese bankers with those qualifications. But to find a top-notch foreign banker who was also fluent in Chinese was like finding a needle in a haystack. I felt that Williams was a godsend; he seemed to fit the bill perfectly. Our only concern was whether he would be tough enough for the job.

I invited Jack Langlois to take the position of chairman. Langlois and I had worked as a team at J.P. Morgan. He was a historian specializing in Chinese history, with a PhD from Princeton University, and became the chair of the history department of Bowdoin College before joining banking. As a historian and linguist, his Chinese was better than most natives', absent any trace of local accent that native Chinese usually bring to the language. At one time, J.P. Morgan's visiting chairman hosted a dinner party in the Great Hall of People in Beijing. It was customary there for dinners to start and finish early. Our guests had all left by about 8:30 p.m. Langlois walked to my table, asking, somewhat exasperatedly, "How come they have all dispersed like birds and animals?" The Chinese idiomatic expression he used, "dispersing like birds and animals," is rarely used even by well-read Chinese, testifying to his mastery of the language.

At the time, Langlois was working at an entity in China set up by Morgan Stanley. While the SDB chairmanship was a nonexecutive position, meaning Langlois did not have to work at the bank full time, he would still need to seek the blessing of Morgan Stanley's senior management. On October 26, we caught a break. That morning, Langlois called to tell me that Philip Purcell, the chairman and CEO of Morgan Stanley, was visiting Beijing and would be meeting with China's central banker at 9:00 a.m. I looked at my watch; it was already 8:45. I hurriedly placed a call to Governor Zhou Xiaochuan, and by luck I was able to reach him. I asked him to help persuade Purcell to allow Langlois to serve as chairman of SDB. I thought it would be hard for Purcell to turn down a direct appeal from China's central banker. Sure enough, three days later, Langlois called to say that Morgan Stanley had given him the permission.

I invited another seasoned J.P. Morgan banker, John T. Olds, to join SDB's board. He had worked at Morgan for more than two decades and was rumored to be a contender for the top job. Then he was tapped by Lee Kuan Yew, Singapore's former prime minister, to become vice chairman and CEO of the Development Bank of Singapore. He was credited with having substantially improved that bank's operation and profitability before he retired. I thought his experience, particularly in managing an Asian bank, would be immensely valuable to us.

We also invited Frank Newman and Mike O'Hanlon to join the board. Both were veteran bankers and had served on the board of KFB with Newbridge. They were among the most conscientious and devoted directors we knew, never hesitant to dispense candid and considered views. Newman was the former chairman and CEO of Bankers Trust. O'Hanlon was head of the financial services group of Lehman Brothers for Asia. We knew both men to be outstanding bankers, trustworthy and diligent in discharging their duties.

My colleagues Dan Carroll, Au Ngai, and Tim Dattels, who had joined Newbridge as a partner after retiring from the position of Asia head of Goldman Sachs, an investment bank, rounded out our lineup for the SDB board. Overall, it was a strong group. In deference to the wishes of Shenzhen government, I did not immediately join the board. We had an understanding that I would remain in the shadows until things calmed down. But this did not prevent me from interacting and working with board members and management; people knew I represented the controlling shareholder. I felt a deep sense of responsibility and ownership for this investment and remained involved in major decisions on an almost daily basis, though I carried no official position or title within the bank.

The heads of both the PBOC and the CBRC gave us strong support; clearly, they expected us to improve the operation of the bank. On December 23, 2004, a week before we closed the deal, I spoke with the CBRC chairman, Liu Mingkang. He told me that both the CBRC and the State Council were well aware of the problems within SDB. Therefore, he assured us, Newbridge "will receive credit after making all the efforts to turn it around."

He continued, as I noted in my journal:

> Your team has exposed many problems inside SDB. The PBOC's Financial Stability Bureau will provide support with liquidity through on-lending if you ever need it. Secondly, please report to us if there are difficulties and problems at the head office level. Thirdly, the CBRC and the PBOC will coordinate with the CSRC to provide policy support as yours is a listed company.

By "policy support," Liu meant such things as the rights issue, which required the approvals of both the CSRC and the CBRC.

*And fourthly, with regard to licensing and permission, we will use soft
and new standards, reasonable and considerate, during SDB's transition
period. The PBOC and the CBRC have reached consensus between
us to provide help. You have a national franchise headquartered in
a border area so you can take advantage of the resources from Hong
Kong and your bank will be a bridgehead. You should have a new
team, independent and capable with new people and new thinking.
Please also cooperate with the city government.*

Under Chinese banking regulations, opening new branches and
offering new products required licenses and permission. If a bank did not
have adequate capital or was poorly managed, regulators usually would
not allow the bank to expand. But in our case, the regulators were well
aware of SDB's problems and eager to see it turned around. Liu's "soft
standard"—in other words, regulatory forbearance—would give the bank
a chance to grow out of its problems.

Liu Mingkang spoke English fluently. At this moment, he suddenly
switched to English.

"Tell your partners," he said, "that in case of any emergency, like
a bank run, the PBOC and the CBRC will be on standby to provide
liquidity support." The PBOC and the CBRC would also give SDB time
to improve itself before enforcing certain banking rules, such as meeting
with the required capital ratio, he assured me. SDB would not be alone in
this regard, "We will also provide a package for Guangdong Development
Bank to bail it out," he added. "When we clean up the banking system,
everyone will be happy."

I understood that he spoke these words in English so that I could pass
his message accurately to my American partners. Crucially, SDB would
be exempt from certain public disclosure requirements. Liu knew that
SDB's true NPL level was far greater, and its capital ratio far lower, than
stated in its published reports. But disclosing the true numbers would
have forced the hands of regulators to impose penalties on SDB, and the
bank was already too weak to absorb any further blows. Real transpar-
ency might also cause a panic in the market and among SDB customers.
However, Liu told me, he did not want the bank's management to know
about this implicit regulatory support, lest they lose any sense of urgency
to fix things.

Liu's words amply reflected how banking regulators were worried about SDB's health. While they would do what they could to help the bank, they counted on us to save it and to turn it around. The regulators had a daunting task to reform and clean up the entire banking system, as Liu pointed out; SDB was not alone in being troubled. But it was one of the weakest chinks in the armor of China's banking system—if it failed, the contagion effect would be huge and unimaginable. Saving it would be an uphill battle neither Newbridge Capital nor the regulators could afford to lose.

It was 2005 and the beginning of a new era for China's banking reform—SDB led the way, to become the first and only national bank whose control was taken over by an American firm. In the next two years, four of the five biggest Chinese banks would complete their restructurings, bring in foreign strategic investors from among the best banks in the world, and go public on Hong Kong's stock exchange.

The author speaking on January 25, 2005, at the closing ceremony of Newbridge Capital's investment in SDB.

Jack Langlois, the first chairman of SDB under Newbridge's control, speaking on January 25, 2005.

Zhou Xiaochuan, governor of the People's Bank of China (the central bank), speaking at the closing ceremony of Newbridge Capital's taking control of SDB.

Zhou Xiaochuan and Dick Blum, co-chairman of Newbridge Capital, shaking hands after Zhou's speech.

David Bonderman, chairman and founding partner of TPG and co-chairman of Newbridge Capital, and the author at the closing ceremony.

At the closing ceremony, from left to right, Au Ngai, Jack Langlois, Jeffrey Williams, David Bonderman, Dick Blum, Zhou Xiaochuan, and Dan Carroll, with others. The author is second from the right.

Chapter 14

Righting the Ship

There was much anticipation, excitement, and anxiety at SDB as the new year dawned. We had a new board, new chairman, and new president. More changes were expected. But what? How? Who would be affected? Nobody knew. The job to transform SDB fell chiefly on the shoulders of our new president, Jeffrey Williams.

The wheels started coming off almost immediately. In early February, barely a month after we had taken control of the bank, I confronted Williams with a stark question: "Jeff, why did the bank's deposits fall by 3 billion yuan in the month of January?"

I had been reviewing the financial summaries with him and I was alarmed by the sharp drop over the few weeks the bank had been under our control. SDB was severely short of capital, and the weight of its non-performing loans meant that it had limited capacity to meet a large-scale withdrawal of deposits by customers. If this trend continued, the bank might literally run out of cash to meet its obligations. And that, of course, was how a bank failed.

Williams was unruffled. The large drop in deposits was a seasonal phenomenon, he explained, typical at the beginning of the year, when economic activity slowed down due to holidays. He assured us that deposits would return in the next month.

They did not. In February, SDB's total deposits continued to drop, and they dropped precipitously. Now 7 billion yuan of deposits had gone out the door since January 1. I was truly worried. I had a long talk with Williams, reminding him that SDB had inadequate capital and depended on these deposits for liquidity and for survival. He shouldn't have needed reminding. Even though the PBOC had assured us that it would provide SDB with funding support if necessary, we all knew that once we got hooked on life support from the central bank, we would not be able to wean off from it; the market would know we were in trouble, and nobody would want to bank with us.

Still, Williams seemed unfazed. He was confident that the bank would be able to bring back the lost deposits in the months ahead. I wanted to believe him, but I cautioned him that 10 billion yuan in lost deposits would be the red line, beyond which the bank would be in grave danger.

On March 2, I learned that the cumulative deposit drop had reached 10.3 billion yuan. On that same day, Dan Carroll called me to say that he was losing sleep over our B&B loan, which was still hanging over our heads. The entire Newbridge team was gripped with fear. After all we had done to wrest control of it, the bank teetered on the brink of failure only two months into our tenure as stewards.

My colleagues and I had a long conversation with Williams, who was now the only one showing little sign of worry. We did not know whether to take comfort from his calm or be even more concerned that our bank's president did not seem to appreciate the gravity of the situation.

I challenged Jeff, trying to convey as clearly as possible the sense of a crisis. I asked who in his team was specifically in charge of branches and deposits.

"I am in charge," he replied. "No one else."

I was astonished. That was his answer, in its entirety: "I am in charge." Here was a national bank with some 200 branches scattered across a vast country, the size of the United States, with more than 7,000 employees. How could he manage the bank without a team?

"I don't know who to trust," Williams explained, "because everyone was telling me that everyone else was a crook."

Well, that much was true. The bank had been so poorly managed that there were not only incompetent managers in key positions, but also some crooks as well—people who had connived with dubious customers to siphon money from the bank, or had made loans in exchange for personal favors. As a newcomer and a foreigner, how could Williams possibly tell who was good or bad? How could he know whom to trust? If he appointed one of the crooks to a senior position, I knew, there were consequences beyond the immediate damage that person could do themselves; rotten apples tended to attract rotten apples, pushing good people away. Indeed, we had discovered that nepotism was rampant within the bank and was part of the old culture.

Despite this, there was no way that one foreign president could manage a Chinese bank without the help of a senior team. I had warned Williams not to breach the 10 billion yuan red line, but by early March, the bank had raced past that marker already. My partners and I decided we had to take immediate action to rectify the situation and find a way to stop the bleeding.

Williams had worked with some of the best banks in the world and came to us with impeccable credentials. But he had never led a bad bank before. In his world, bankers followed established policies and processes. SDB was more like a listing ship that needed to be righted before it sank, with a panicked and undisciplined crew running around the deck. If it was his first time in this situation, even a well-trained captain might not have known what to do.

We needed someone with experience in turning around a bad bank. Or we risked running aground.

★ ★ ★

I met with Chairman Liu Mingkang of the CBRC again, to consult with him about the grave situation we found ourselves in. I suggested that we might have to find a new head for the bank, someone with experience in turning around troubled institutions. Liu was quite alarmed.

"Weijian," he said gravely, "we have never had a foreigner as president of a national bank in China since the founding of the People's Republic. The appointment of a foreign president was such a high-profile move that it had required special approval from the State Council, and then we announced it to the world."

"Now," said Liu, "barely two months into his tenure, you want to replace him? No. That wouldn't look good at all."

I knew it wouldn't; we certainly shared his concern about making such a drastic change so soon after taking control. But the alternative, in my view, would be much worse: "We are in a liquidity crisis already," I said. "What if the bank fails?"

"No. You certainly can't let that happen. We can't allow that to happen," he repeated, emphatically. "Okay. We will allow you to replace the CEO if you determine it's necessary. But you have to do it in a very low-key way, without alarming the market."

The partners of Newbridge had a call to discuss what to do. We decided that I should step in to help Williams. Bonderman had some doubts. "He will probably resign," Bonderman cautioned, as usually a CEO would not want someone else to be in the driver's seat, or even to be second-guessed. We were all reminded of Chairman Liu's warning to keep the change low key, so Bonderman's concern was well founded. But we concluded there was no other choice but to take that risk.

I held no official position in the bank, but I thought an unusual situation required unusual measures. First, I asked Williams to appoint me as senior advisor to the president.

The next thing I did was to form a new top decision-making body. Officially, such powers in any Chinese bank rested with the President's Council, consisting of a bank president's direct reports. But Williams was not using his council to make decisions, presumably because he felt he could not trust all its members. Like Williams, I had little idea whom to trust among the senior executives within the bank, although, being Chinese, I was more plugged in to the team than he was. Staff members were more willing to confide in me, some for personal reasons and others for the good of the bank. I needed to find senior managers we could trust to help implement the major changes necessary to put SDB on the right track. How would we go about finding them?

I had suggested back in January that we create a core group of well-respected and capable senior managers to support the CEO, selected by using a 360-degree performance review process. Under this process, each individual would be reviewed and evaluated by the people around them, including superiors, subordinates, and colleagues. Evaluations were kept anonymous to ensure as much fairness and honesty as possible. My advice had been ignored. Now I pushed out the performance review program with a tight timetable. In two weeks, we received all the results.

This had never been done at the bank before. Judging by the results, employees welcomed the opportunity and participated fully. I was given summaries for all the evaluees in a large spreadsheet. The ratings were telling. Managers of branches with good-quality assets were generally rated higher than those of poor-performing branches. The review system might not have been perfect, but it served its purpose: The bank's employees knew who the best managers were, and given the right outlet, they were willing to tell us.

I picked a few people who had received the highest ratings to be candidates for the core management team. I did a reference check on each one of them by talking with other high-rated senior officers of the bank. And then I interviewed all the candidates. This process produced a few finalists whom I invited to become members of the new leadership team. After consulting with Williams, I labeled this new body the Strategy and Policy Committee (SPC).

The members were Liu Baorui, who was already a deputy president of the bank; Hu Yuefei, who until then had been the head of the Guangzhou branch in Guangdong Province; and Zhou Li, who had headed the Hangzhou branch in Zhejiang Province. I thought the three of them, plus Williams and myself, would make a good committee. I would effectively promote Hu Yuefei and Zhou Li, two provincial branch heads, into the top management team, bypassing whatever processes the bank had at the time. It was far from the ordinary way of doing things—but it was an extraordinary situation that required extraordinary measures.

Chinese banks were organized like a federation of regional fiefdoms. There was a major branch in the capital of each province, which functioned like a regional headquarters controlling a number of sub-branches throughout the province. Guangzhou is the capital of Guangdong Province, with a population of roughly 100 million people. Hangzhou, near Shanghai, is the capital of Zhejiang Province, with a population of 55 million. These were massive markets, not only by population but also by their relative affluence compared to the rest of China.

SDB had a high ratio of NPLs in general. But the bad loans were unevenly spread: Some branches had much higher NPL levels than others. Despite the poor risk management system of the bank as a whole, we found that the balance sheets of some branches, such as Guangzhou, which Hu Yuefei headed, and Hangzhou, where Zhou Li was in charge, were relatively clean and the asset quality generally good. In contrast, about 70–80% of SDB's nonperforming loans were concentrated in the Shenzhen area, even though the loans booked by the Shenzhen branch accounted for only 30% of the bank's total. The Shenzhen branch was a disaster zone.

What caused this uneven spread of NPLs? I had read an early-2000s study by the economist Yi Gang, then an advisor to the central bank governor (Yi himself would become the country's central banker in March 2018), who analyzed the credit quality in China's banking system across different geographical regions. I knew Yi quite well. He was an accomplished scholar with a PhD from the University of Illinois and a tenured professorship at Indiana University. When I was teaching at the Wharton School, I was the chief editor of the *China Economic Review*, an academic journal, and Yi was a frequent contributor.

In his study, Yi divided China into two regions, rich and poor. Then he divided these regions into two buckets, those with a historically good credit culture and those without. "Culture," by definition, is a set of beliefs and practices. For a bank, a good credit culture means that it has a system to ensure that lending decisions are made on the basis of the ability of the borrowers to pay back the loan, or the credit quality of the borrower, not on the basis of personal relationships or government policies. Similarly, people in some regions are more inclined to pay back their debt obligations than in some other regions. It may have to do with wealth or economic conditions—loan default rate rises in a recession—but not entirely. Those regions with a low debt default rate on average are considered to have a good credit culture.

Yi used a two-dimensional graph to show the regions' four resulting buckets: (1) rich with good credit culture; (2) rich with weak credit culture; (3) poor with good credit culture; and (4) poor with weak credit culture. His analysis contradicted the conventional wisdom that the credit culture in poor regions was worse because firms there were less capable of paying back their loans. He showed that in a poor province like Gansu, the loan quality of the banks was generally quite good, whereas in a rich province like Guangdong, the loan quality was surprisingly poor. He attributed this pattern to the credit culture unique to each region.

I thought Yi Gang's analysis was insightful and very useful for banks as they considered how to manage their risks in different regions. But it also gave me the impression that the loan quality might be determined more by where a bank's branches were located than the soundness of its risk management system. If a branch was located in a zone of weak credit culture, such as Guangdong and Henan, it might be difficult for such a branch to keep a clean balance sheet. Bad credit culture was also contagious. I remembered *The Economist* had reported that during the Asian Financial Crisis, banks from South Korea to Indonesia faced borrowers who engaged in "strategic default"—they defaulted on their bank loans even though they were capable of paying, because they saw their peers defaulting left and right and they felt unfairly disadvantaged if they alone had to honor their debt.

Still, what I discovered surprised me. The loan quality of our Guangzhou branch was generally quite good, even though the province was considered to have a weak credit culture. The loan quality of the Shanghai branch, meanwhile, was surprisingly poor—although Shanghai, the country's largest and most developed metropolitan area, was generally considered a good credit risk.

What was going on here? Was Yi Gang wrong in his analysis? No. His analysis was robust, and his findings were based on actual data. But why was SDB different?

Clearly the managers of the SDB branches had made the difference, good or bad. If the branch manager was good, the branch booked good loans and avoided bad ones—even in a bad market. Conversely, a bad branch manager produced bad loans, even in a good market. Yes, some markets were more difficult than others, but a good branch manager could overcome the difficulties in a region known for its bad credit culture, whereas a bad one could generate a mountain of bad loans even in the best credit environment.

The other factor here was SDB's organizational structure. The provincial branches of Chinese banks, including SDB, operated with their own lending standards and made their own credit decisions—with a wide range of results. From the point of view of international best practices, the whole bank should have a uniform credit underwriting standard. This situation was unthinkable.

So it was quite remarkable to learn that Hu Yuefei had been able to build a loan book of good quality in Guangdong, a region of weak credit culture. Likewise, Zhou Li had done well in Zhejiang.

Both were relatively senior: In addition to carrying the title of branch manager, they were "assistant presidents" of the bank, just one level below deputy president. I thought promoting them would not be controversial, and that the two men might bring a breath of fresh air to the senior ranks.

I spoke with Hu Yuefei and Zhou Li separately about forming the SPC and including them as members. Even though neither I nor Williams had the power to give them new titles (those could only be conferred by the board and approved by the CBRC), I proposed to immediately put Hu in charge of all branches, a job usually held by a senior vice president. This meant he would be able to drive all branch managers nationwide to focus on, say, deposit collecting. In view of the liquidity crisis faced by the bank, this would be the most important position other than the presidency itself.

I knew he would be up to the job. He was 43 years old but had been in banking for 26 years, since he was 17. He had joined SDB 15 years ago, when he was 28, and rose through the ranks.

To my great surprise, Hu told me he would rather keep his old job than move to the head office. He didn't think he was capable enough to take on more responsibilities, he said. He felt comfortable managing one provincial branch. I listened carefully, but I was in a hurry to put a team together and he was one of the best candidates I had. I wasn't going to take *no* for an answer.

"I will give you 24 hours to think about my proposal," I said to him. "If you decide not to accept my offer, then you should quit. You won't be going back to the Guangzhou branch anyhow."

It was his turn to be surprised. "Why?" he asked. "Why couldn't I continue to manage my branch? Have I not done a good job?"

It was a reasonable question. In my answer, I tried something that might have been a bit of a stretch. "Well, Napoleon once said, 'a soldier who doesn't want to become a general can't be a good soldier.' If you don't want to be promoted, you can't be that good. If so, you'd better just quit."

I wasn't 100% sure if Napoleon had actually said those words, though I vaguely remembered having read it somewhere. I thought it was a good line to use anyhow, for that particular occasion. I meant to be provocative. In any case, he came back the next day and accepted my offer. It seems that Napoleon Bonaparte's words will have a more lasting impact than his wars.

★ ★ ★

It was easier to persuade Zhou Li to take a new job. He was 2 years older than Hu and had been with the bank for 13 years. With a PhD in finance, he used to teach at Nankai University, one of the most prestigious in the country. He was smart and capable. We offered him control of the Shenzhen branch.

Shenzhen was not an ordinary branch for SDB; it was the largest in the entire bank, home of 89 of the bank's 225 branches and sub-branches. We could not possibly turn around SDB without turning around the Shenzhen branch. Like Hu Yaofei, Zhou Li initially expressed reluctance about my offer. His feelings, like Hu's, were perhaps a product of the Chinese cultural norm of showing modesty and humility. (This cultural norm was so strong that, historically, a number of usurpers of the imperial throne had had to go through the motions of rejecting the pleas of top officials at least three times before accepting the crown.) I had little patience now for such niceties, which in my view bordered on hypocrisy, and I went with the more American style of being direct and blunt. I told him he could either take my offer or quit.

He took it.

We held the first Strategy and Policy Committee meeting on March 19, in the 32nd floor conference room at the bank's headquarters. There were five of us there—Jeff Williams, Liu Baorui, Hu Yuefei, Zhou Li, and me. And there were three agenda items:

1. Form the SPC.
2. Create a new organizational structure.
3. Change management at various branches.

From the minutes of the meeting:

> *President Williams needs broad and strong support in the decision-making process, but the current President Council has too many members, which makes it very difficult to make any decisions. Based on the*

ratings, Williams invites Liu Baorui, Hu Yuefei, and Zhou Li to join himself to form a "Strategy and Policy Committee." The Committee is to serve as an advisory body to aid the president in setting strategy and policy.

In effect, from now on all the major decisions would be made by the SPC. In that first meeting, we resolved that Liu Baorui would be put in charge of retail banking, a new area for the bank, which had historically focused on the business of corporate customers; that Hu was to be responsible for corporate banking and branches outside Shenzhen; and that the Shenzhen branch would fall under the responsibility of Zhou Li.

The minutes recorded the conundrum we were in:

Mr. Shan points out that nepotism cannot be tolerated as it seriously undermines the management and business. All have observed that the current situation at SDB is that

- *there are few good branch managers*
- *there are few good management teams*
- *business at each branch is still vulnerable*

At my suggestion, the committee resolved to replace the managers of the vast majority of SDB's provincial branches. We also made personnel decisions for key posts at the head office, including the appointment of Chen Rong as head of risk management. Only 37 years old, she had just returned about half a year earlier from studying in England, but she had been the chief of the credit department.

I had kept an initial list of people to be removed and installed, based on ratings from the review process. But we made our changes after thorough discussion and consideration among the committee members, based on input from old-timers who had good knowledge of the individual candidates and market conditions in particular regions.

The changes represented a major restructuring of the bank's management. The move was so bold that some of my own colleagues expressed reservations. They were worried that the bank was too weak to withstand what was effectively a kind of shock therapy. Shouldn't we make changes more gradually?

I didn't think so. I thought the bank was like a gravely ill patient who required immediate and major surgery. The longer we waited, the greater the danger to the health of our patient. The bank was already suffering from bleeding of its deposits. I felt that the surgery required was like a bone marrow transplant, aimed at generating new blood cells to replace the diseased ones. The benefits, in my view, would far outweigh the risks.

Sure enough, the announcement of the management changes landed like a bombshell. Some of those who lost their jobs because of poor performance would not go without a fight. Tensions grew so high that some of my colleagues received threats of bodily harm. Our U.S. partners were worried enough to propose that our own team temporarily relocate to the United States. Some of our advisors offered to engage private security services for me and my colleagues. Others warned me to stay away from Shenzhen. I thanked them for their concern, but I saw no need to take those measures.

I also did not think we could turn around the bank if we had to hide. I was convinced we were doing the right things, all for the good of the bank and the employees. We had let the results of the reviews guide our personnel decisions. None of those decisions were personal. I wasn't overly worried about safety, but I did take the precaution of having a driver take me across the border between Shenzhen and Hong Kong instead of driving myself, as I had done before.

In general, the changes were positively received and supported by the staff. The reason was simple: There had been a culture of nepotism at the bank, with promotions depending more on personal relationships than performance. By promoting people based on the results of an objective review and evaluation system, we had sent a strong message to the rank and file that future advancement would be given on the basis of merit and performance, nothing else. I believe this approach significantly boosted morale.

In March, I was invited by Chen Rong, the newly promoted head of risk management, to give a talk at a first-ever employee training session on risk management. I began by asking those in the audience whether they considered SDB to be among the best banks in China.

Not a single hand went up. Remarkable, I thought—and not in a good way. Not one among the 30 or so employees in the room thought SDB was one of the best banks in the country.

My next question: "Then how many of you think our bank ranks in the middle in China's banking industry?"

A few scattered hands this time. "*Hmm*, less than 20%," I said when I finished counting.

"Then how many think that our bank ranks at the bottom?"

More hands. I looked around and said, "Okay. A majority."

Some people hadn't raised their hands at all.

"So, some of you think our bank doesn't exist?" I quipped, eliciting laughter.

The audience seemed relaxed. From my experiences, particularly as a professor at Wharton, I knew that an American audience was typically lively and engaged, whereas an audience in Asia was usually passive and reserved. From their lit-up faces and reactions, I could tell this group was attentive, almost enthusiastic.

"It seems," I continued, "that most of you think our bank ranks at the bottom. This is a sad thing. We all work for SDB, yet we don't think it's a good bank. We need to think about why this is the case."

I pointed out that not all the branches were bad, and that a number of them, including Hangzhou, Guangzhou, and Nanjing, had produced good-quality assets. I shared my view that the difference was often made by the manager, not by the market or by other external factors.

"Where there is a will, there is a way," I said. "If a person wants to do something well, then he can. However, to do it well, we must have a good incentive system, a system of reward and penalty."

I went on to talk about the importance of a good system. It was well-known that China's state-owned system had produced inefficient firms. But then I provided some counterexamples. I cited some well-known state-owned companies that were well-run, such as Singapore Airlines and Baosteel, and some private companies, such as Enron, that had gotten themselves into trouble.

Then I challenged the audience rhetorically. "There are good state-owned companies and there are bad private ones. If so, why do we still want a market economy?"

My audience seemed puzzled. Judging by their silence, I could tell none of them had ever considered this issue. The answer was, of course, not obvious.

"I've given much thought to this," I went on. "China had for a few thousand years the system of absolute monarchy, which we all know was a bad system. However, in some periods in its history, China was strong and prosperous. But why do we think absolute monarchy is no good?"

A woman raised her hand. "Because there were too many bad emperors?" she ventured.

"That's right," I said. "It is because such a system didn't have a way of getting rid of bad emperors who brought disasters to the country."

"Similarly, the state-owned system allows bad companies to continue to exist," I continued. "The market economy is better because it has a mechanism to get rid of bad apples."

Now I got to my point: "So how do we build a good bank? Taking a page from history, we also need a system to reward the good and punish the bad. We need to be able to bring in good people and to get rid of rotten apples. Good performance must be rewarded, and bad must be punished. If this system becomes part of our culture, there is no reason for our bank not to become a good one."

Surveying the room, I saw nods all around, and a few smiles. I appealed to the employees' sense of ownership.

"Whose bank is this?" I asked. "Is it mine? Or is it yours? … This bank belongs to you, me, shareholders, employees, and customers, including all depositors and all borrowers. We have several thousand employees. Everyone should consider themselves the owner. If everyone thinks that way, the bank will make progress and your personal career will advance. For a company to sustain itself, a sense of ownership is a must."

I have always believed strongly in the power of this sense of ownership, and the pride that comes with it. If someone considers a job their own, then that employee will be far more likely to do it well. A sense of ownership is critically important. I was blunt in addressing what I considered the major problems with the bank—poor management and low morale chief among them. We had to change that culture. I finished my talk with a few words about the importance of having a strong credit culture—making lending decisions based on the credit quality of the borrower. "It goes right to the life and death of our bank," I said. "If all of us can help build a good credit culture, our bank will no doubt become a good bank."

When I finished, the audience applauded. I could tell that they genuinely approved of my words.

A transcript of my remarks appeared in an internal publication of the bank. A number of employees told me later that they understood from reading or hearing about what I'd said that Newbridge, as the controlling shareholder, was focused on making SDB a good bank, and making their career a success. That indeed had been my message to the staff.

* * *

One major change that Williams initiated was the centralization of decision making in the credit review committee at the head office. This would ensure uniformity in the quality of loans. Such a move took the credit approval power away from branch managers, who were now responsible only for marketing the bank's products and generating loan applications. There was a credit officer in each of the branches, but the credit officer answered directly to the chief credit officer in the head office, not to the branch manager. This system would prevent branches from making loans based on personal relationships; it also freed frontline managers to maximize their marketing efforts for loan volume.

Williams coined a message, like a campaign slogan: One Bank. The idea was that SDB was no longer a federation of fiefdoms, each with its own rules and standards—or, more often than not, a lack thereof. SDB would now have one single system of centralized controls and standards in loan decisions, human resources, risk management, and compensation. The One Bank message quickly caught on and soon all employees understood the term, and the new direction.

Meanwhile, the appointment of Hu Yuefei to lead corporate banking and branches was reaping fruit at a pace we could hardly have anticipated. The turnaround was stunning—nowhere more so than in stemming the outflow of deposits. Within a few months, deposits were flowing in and our liquidity situation began to improve. In a mere two months, Hu brought back nearly the entire volume of deposits lost in the first quarter of 2005—much of that thanks to a simple reward system he put in place for the staff: bonuses tied to the amount of deposits collected. Some even convinced their own relatives to deposit their money with SDB. From that point on, SDB never looked back. The bank continued to grow its deposit base and its loan book for as long as we owned it.

★ ★ ★

The major problem that remained was the mountain of nonperforming loans. Any branch with a large number of NPLs could not make new loans without first collecting the unpaid loans. The bank could not possibly grow with such a heavy burden.

There was no question that Hu had helped avert disaster on the deposit side, but the financial condition of the bank still looked dire enough that on April 5, John Olds, the veteran J.P. Morgan banker, resigned as a director of the board. He said he was too concerned about the risks, and I couldn't blame him. Olds had enjoyed an illustrious career in banking and probably didn't want to tarnish his own reputation as a board member for a troubled institution. Fortunately, his resignation didn't affect the normal functioning of the board, as he had been a relatively hands-off director. In contrast, other independent directors such as Jack Langlois, Frank Newman, and Mike O'Hanlon were working closely with the Newbridge team and management to help set the bank on the right course. Olds' resignation underscored the dire straits SDB was in, but the rest of the board was undeterred in the resolve to lead a successful transformation.

There was a tried-and-tested model for restructuring a problem bank. Effectively it called for the creation of two separate institutions: a "good bank," which would manage all the performing loans, and a "bad bank" housing all the bad loans. The good bank, free from legacy issues, could grow again. The bad bank would focus on recovering whatever it could of the outstanding debt. Bonderman pioneered this model when he and his partners acquired American Savings Bank. Newbridge's acquisition of Korea First Bank also followed this model. In the cases of both ASB and KFB, the government took the ownership and responsibility for the bad bank. In the case of SDB, Newbridge inherited a bank that mixed good and bad assets together.

We decided we would create a Special Asset Department—effectively, a "bad bank" within SDB. All of SDB's nonperforming loans would be transferred to this internal bad bank. Freed of bad loans, the rest of the bank would be able to focus on managing and originating good loans.

The internal bad bank would be responsible for collecting and resolving all nonperforming loans. And I knew exactly whom I wanted to lead it.

We held an SPC meeting on April 1. At my recommendation, the committee decided to hire Wang Ji, the former chief negotiator for the Shenzhen city government, as head of the Special Assets Department.

Despite all that had happened to him, Wang remained a man of great energy and drive. Before being ousted as president of Shenzhen Commercial Bank, he had been widely credited with singlehandedly cleaning up its mountain of bad loans. What he lacked in tactics, he made up for with single-minded persistence and an uncompromising sense of right and wrong. He would not let anyone get away with pillaging his bank. Collecting debts required courage because of the personal risks involved; some debtors could take extreme measures to resist collection. Wang was fearless. He was also highly motivated: While he had ultimately been exonerated, the corruption investigation that had ensnared him had tarnished his reputation and cost him his job. He was eager to get back into action. He was, I thought, the perfect person to be our "NPL tsar."

Chapter 15

Bank Repairman

The Strategy and Policy Committee (SPC) made some major decisions to overhaul the operation of the bank, from key personnel changes to the system to deal with nonperforming loan problems. But I still felt that much more needed to be done to right this ship. My colleagues agreed. We needed more changes, and we needed them faster than Jeff Williams was delivering. I wrote to him on April 9 summarizing a long list of immediate and major issues as I saw them:

Now that we have revamped the management for our branch network and have put the members of the policy committee in charge of retail, corporate banking, and branches, we have a much more stable organization. It seems that these changes have significantly lifted the morale of our staff throughout the bank. With the replacement of the Beijing branch manager next week, we will have dropped the other shoe for our branches, by and large, which will also remove any concerns in the minds of those branch managers largely unaffected by the revamp.

The first-quarter results show that we have outperformed the plan for P&L [profit and loss], but we have significantly underperformed the plan for the balance sheet which shrank with both loans and deposits short of plan by about 4 billion yuan. This, coupled with the replacement of 3 billion yuan or so by the more expensive postal deposits, will have a severe effect on net interest income in the quarters ahead unless we can replenish the lost deposits and grow the loan book within a relatively short period of time.

The postal deposits were an issue. In China at the time, interest rates, both for deposits and loans, were controlled by the central bank. But large corporations were permitted to negotiate a higher interest rate for their deposits than ordinary retail clients'. China's vast postal system served as a savings collector, gathering deposits from its customers and depositing the funds with banks. Because of its enormous size and a network that touched every corner of the country, it had outsized bargaining power to get the highest interest rates available. These high-interest-rate deposits brought down the profit margin of the bank, of course. I continued:

Without so much as a clearly articulated strategic plan, our consensus view, as captured in the Newbridge/Morgan Stanley model, is to aggressively grow the retail business while maintaining the growth on the corporate side. We hope to have a balanced corporate and retail portfolio in about three years' time. Of course, we assume that we will be able to recapitalize the bank with a rights issue and/or other recapitalization plans, and that the risk control of the bank will be brought in line with international best practices.

To achieve the objective of meeting the plan for the year and achieving our strategic targets in the years ahead, I think we still have a number of very urgent tasks on our hands which cannot wait, or we risk not being able to deliver what our plans call for.

I went on to enumerate those tasks. We needed the bank branches and central credit control team to align on credit policies and procedures, I reminded him, and we needed better finance policies to incentivize staff, in particular to design incentives for NPL recovery. "Many of the departments at the head office remain disorganized," I wrote. "You let go of [a senior risk officer]. To date, there is no replacement." Branch managers didn't know whom to turn to for risk policies; on this and other matters I wrote that "I regret to say that our decisions are ignored, which may have broader repercussions." We needed to launch the retail side of the business, reduce redundancies in headcount, and fill critical senior positions, including our chief financial officer, chief technology officer, and head of human resources.

It was a pretty withering message, and I finished it this way: "These are all on my mind. Since we didn't have enough time to cover these issues in the last two policy committee meetings, I thought I would write them down to see if we could get things moving without having to wait for the weekly meetings."

Williams responded as he often did—quietly and calmly; he did not push back against any of the main points. I was unsure, however, whether he felt the same urgency that I did or if he would take decisive actions to address the issues I raised.

On April 15, we held another SPC meeting, welcoming a new member, Wang Ji, the NPL tsar. Although we covered many urgent issues, I left the meeting frustrated; no specific action plan or timetable was put forward. I wrote a long memo to the attendees on the following day, again highlighting the areas where we were falling short, and the urgent tasks ahead:

> *I feel that although we spent a lot of time discussing how to better manage NPLs yesterday, we didn't achieve much, and we didn't reach a consensus which can be implemented for the long term. The key problem remains nonconvergence of thinking, unclear plans, and lack of thorough analysis of feasibility of the plans. Therefore, we didn't make marked progress from the last SPC meeting. The result is that we continued to endlessly debate and rationalize whether we should do this or that, issues we had already reached conclusions on before. However, we failed to reach a consensus view on basic principles. Therefore, with regard to how to take the next step, other than the clarity with Mr. Zhou Li's plan to carve NPLs out of Shenzhen branches, which plan comes from the selfish point of view of Shenzhen branch (you can't blame him as he is in charge of Shenzhen branches) and which is almost irrelevant to the master plan, the rest of the overall master plan remains murky.*

The two most urgent priorities for me were that the bank's branches had to grow new businesses, and that there needed to be an immediate carve-out of NPLs from the Shenzhen branches. Meanwhile, the head office should inject an initial batch of funds into Shenzhen in partial compensation for the carved-out NPLs to enable the Shenzhen branch to grow.

With detailed instructions given to Williams and the rest of the senior management, I was stepping deeper and deeper into effectively running the bank. As I did so, my relationship with Williams was fraying. I was frustrated that critical things were not being done or not done fast enough. He no doubt felt irritated with me for usurping his power. Both of us wanted a change.

A week later, he threw down the gauntlet. On April 21, 2005, Bonderman called from Seoul to tell me that he and Dan Carroll had received a letter from Williams requesting that Newbridge stop interfering in the management of SDB. His request came with a threat: If our "interference" continued, he would not sign off on the audited financial statements for 2004. As a publicly listed company, there would of course be severe repercussions if the bank could not issue its audited annual report on time.

As it happened, Blum, Bonderman, Carroll, and I had met in Seoul the day before Bonderman received Williams's letter. During the meeting, we had reached a decision to find a new CEO for SDB. I was mindful of the warning voiced by Chairman Liu of the CBRC not to rattle the market, so I suggested that we still keep Williams on board, and have him report to the new CEO. The letter from Williams that arrived the following day ended any debate. We decided to follow through with our decision.

The next day, April 22, a meeting was held in Newbridge's office in Hong Kong with Langlois and several senior figures from Newbridge— Bonderman, Blum, Carroll, Au Ngai, Ricky Lau, and me. We discussed how we might go about finding a new CEO for SDB. We had to tread with great care. The appointment would require regulatory approvals in addition to support by the board. We had to find someone with impeccable credentials; regulatory rejection at this stage would be calamitous. More importantly, any candidate, no matter how experienced and credentialed in banking, might not work out at SDB because the bank was such a mess. This had been a lesson of Williams's tenure. Where could we find such a candidate, who we felt confident would succeed? I had a feeling of *déjà vu*. We had faced exactly the same dilemma when we decided to replace the CEO of KFB one year after our investment.

Langlois came up with what we all thought was a brilliant idea. He suggested we invite Frank Newman to be an interim CEO while we searched for a permanent one.

We all knew Newman well. He was an experienced and respected banker and CEO. He had led or helped lead the turnaround of several troubled financial institutions, including Wells Fargo and Bank of America in 1980s, and Bankers Trust in 1990s. He was already a member of SDB's board and had impressed us as a board member of KFB for five years. His only handicap was that he spoke no Chinese. But at this juncture, we thought it was of paramount importance to have a CEO who was an expert in turning around troubled banks; everything else was secondary. And we could not afford another mistake in the choice of CEO and the risk that we might lose credibility with the bank's staff, regulators, and the market itself. Newman was a known factor to us, well respected in the industry, and we felt we could trust him completely.

Bonderman even suggested that not being able to speak Chinese might be an advantage; nobody would bother to ask him for favors, which was how many of the bad loans had come to SDB in the first place. The remaining question was whether Newman himself would be willing to take the job, which would require him and his wife to relocate from New York to Shenzhen, a big move for someone who had never lived outside the United States.

That afternoon, a Newbridge team, together with Langlois, drove to the airport. Serendipitously, Bonderman and Blum were scheduled to fly out a few hours after Newman was expected to arrive in Hong Kong.

Newman must have felt like royalty when he saw all of us there to greet him at the airport. We met in a bare-walled conference room at the Regal Airport Hotel. Blum was the first to broach the subject with him, and we took turns explaining the situation in greater detail. When we had finished, we all stared at him, anxious to hear his response. Newman suddenly felt like our last hope.

Even though he had just landed, Newman was in an immaculate business suit; he usually dressed rather formally. It occurred to me that in all the years I had known him, I had never seen him casually dressed. He wore a wool beret, which gave him a distinguished look. My 13-year-old daughter LeeAnn had once remarked that she could not imagine Uncle Frank ever taking off his suit to go to bed. In her eyes, Newman was a perfect gentleman in both manner and appearance. He conducted business in this way as well. He spoke softly while radiating confidence and seriousness. Eventually his courtly manners and impeccable attire would earn him a reputation as the most English American in China's banking industry.

Newman raised a few questions, but we could tell he was intrigued. He also understood we did not really have an alternative. And then, on the spot—a small miracle for us—he agreed to take the position, but only on interim basis, for six months, until we could find a permanent CEO. He was not ready to commit to a longer term. Still, we were all delighted.

There was actually, in fact, no CEO in a Chinese bank; every bank had a chairman and a president. Usually, the effective CEO of a bank was the president. But we had agreed for Williams to stay in his position as the president, so we had to create a real CEO position above that of the president for the bank. Again, Langlois, who had been nonexecutive chairman, came up with a solution.

"Why don't Frank and I trade places?" he suggested.

It took a few seconds for his idea to sink in. Langlois was proposing that Newman replace him as chairman. It was a beneficent, egoless move, and spoke to the kind of man Jack was. Being the only foreign chairman of a national bank was a big deal, and it had made him instantly famous—everyone in China knew him by his Chinese name, Lan Dezhang. Yet, he was volunteering to give the title away for the good of the institution, without a moment's hesitation.

The best part about the idea was that the switch would, we hoped, be viewed by both the board and regulators as simply a reshuffling of Newbridge-appointed directors. It seemed to be the easiest way to install Newman as the chief executive.

It was, of course, not so simple, because it was not lost on the regulators and on the board that the new chairman would be an *executive* one, managing the bank on a daily basis, whereas Langlois had served as a more hands-off nonexecutive chairman. While the board quickly approved Newman's appointment with enthusiasm, the approval by the CBRC took some time. Without waiting for the formal regulatory approval to come through, Newman stepped into his position in the second week of May, after China's week-long Labor Day holiday.

The CBRC made an exception to approve Newman's appointment as both chairman and CEO in record time. It helped that he was a former senior government official in the United States, having served as Deputy Secretary of the U.S. Treasury Department during the Clinton administration.

As it turned out, we had chosen well, and we were a little lucky too. Newman wound up liking his new job, and we liked him, so much so that he stayed on as chairman and CEO for as long as we owned SDB.

★ ★ ★

The banking reforms that had begun in China in 1999 involved state-owned banks carving out and moving NPLs from their balance sheets to government-owned asset management companies set up and funded by the Ministry of Finance. Then around 2002–2003, these banks were further recapitalized by the central bank. SDB was not so lucky because the local government, which had owned only an 18% stake, had neither the responsibility nor the financial capability to bail it out of its NPLs. Eventually, none of the different types of financial assistance we had hoped to obtain from the government came through.

Crucially, a one-time cleanup of a bank's bad loans could hardly be considered banking reform. Without changing the way a poorly run bank was managed, new bad loans would be created just as the old ones had been. Returning to the maritime metaphor, a leaky vessel could not be saved by simply bailing out the water. The leaks had to be plugged and sealed.

When answering media questions, Newman self-effacingly called himself a "bank repairman." He knew how to plug leaks, and he knew that SDB required an effective risk management system to stop making bad loans in the first place. This was tricky; lending by definition was a risky business, and there was always a chance that a borrower's ability to repay would be impaired in the future. However, if the credit control was too tight, it risked choking off lending and shrinking the bank's income. A good risk management system would allow a bank to grow its loan book while keeping NPLs to a manageable minimum. Even before Newbridge's takeover, our transition management team had tightened SDB's credit controls, effectively stopping further leakage. But it was a stopgap measure. A system was yet to be set up to allow healthy growth without taking too much risk.

Newman went to work strengthening the centralized credit control system. He brought in Simon Lee from Hong Kong as chief credit officer (CCO). Lee had worked for a number of foreign banks and was an experienced CCO. Every branch of the bank was now subject to a credit limit beyond which approval was required from the credit committee Lee headed. Credit officers in each of the branches reported directly to the CCO, not to branch managers. Crucially, the CEO was not a member of the credit committee and had no authority to make loans. Similarly, the client relations bankers, including branch managers, had no authority to approve loans. On the flipside, credit officers had no authority to originate loans. The separation of origination and credit approval provided checks and balances and an assurance that loans would no longer be made on the basis of personal relationships.

Newman also hired Wang Bomin as the new CFO. Wang had been a senior executive at Taiwan's Taishin Bank. Since Mandarin Chinese is spoken in Taiwan as the rest of China, there was no language barrier for Wang and he quickly fit in with the rest of SDB management.

Meanwhile, the Special Asset Department, headed by Wang Ji, was proving extremely effective at bailing water out of the vessel. Typically, no self-respecting banker would want to work in such a department, because the job of a debt collector was neither exciting nor career-enhancing. Besides, it could sometimes be dangerous; you never knew what a desperate debtor might do. The easiest thing for a bank to do was to dispose of its NPLs through an auction, selling the debts at a substantial discount to asset management companies that specialized in collecting on them. But this was not an option for SDB, at least not in the first few years under Newbridge's control. For one thing, the NPL market was not yet developed in China. Another reason was that SDB could not afford to offer deep discounts; it needed every penny to shore up its weak capital base.

When we set up the Special Asset Department, I proposed that we compensate members of the department, including its head, by paying them a percentage of any money they recovered from the NPLs. My idea was to replicate the incentive compensation system of independent asset management companies. The incentive system proved effective almost immediately. In the first year, Wang's department recovered 1.5 billion yuan (~$185 million) from NPLs that had sat on the books for years. That was equal to a third of the bank's entire capital base at the beginning of 2005. The following year, his department collected more than 2 billion yuan (~$250 million). The Special Asset Department turned, relatively quickly, from a garbage dump to an important profit center for SDB.

Our new system also motivated the staff in their collection efforts. Wang told me a story about a member of his team who had a finger severed in an accident while riding his motorcycle on a debt-collection mission. (His finger was not cut off by the debtor, although, as noted earlier, such risks did exist.) Doctors were able to reattach his finger. But he did not follow strict orders to rest, and was soon on the road again, still on his motorcycle, continuing his work. Eventually he lost his finger because he put too much stress on it, irreparably damaging the healing process. Of course, we didn't want employees risking their health for the work, but the man's actions were emblematic, in an extreme way, of the spirit of that department. Gradually, staff members made so much money by taking a cut of the NPLs they had collected that the department became the envy of the entire bank.

Getting the compensation system right was key to improving the operation, efficiency, and productivity of the bank writ large. SDB had previously used a compensation system that could be described as an "iron rice bowl," a Chinese reference to a fixed pay system regardless of contribution, performance, or how the institution was faring financially. I knew from my own experiences working on a state-owned farm in my youth that such a system only rewarded the lazy.

"You got paid 100,000 [yuan] a year, and 60 percent of it would be called your salary and 40 percent was your bonus," as Newman explained it in an interview with *Institutional Investor* a few years later. "But you got 100,000 a year whether you did a good job or a bad job or didn't even show up at the office."

Under the system we'd set up, SDB's senior executives were now compensated like investment bankers, with their bonuses linked to the profitable growth of the business. For staff frustrated by the favoritism and nepotism of the old system, the idea of pay based on measurable performance had proven very popular. "Most Americans don't realize this is really a capitalist system," Newman told the magazine. "People are highly motivated by incentives."

★ ★ ★

In mature markets, banks should charge higher interest rates for customers considered higher risks, and lower rates for lower-risk customers. The ability to price risk in this way allows banks to do business with customers with weak credit ratings; higher interest rates compensate the bank for possible higher loan losses. That is true in almost all markets in the world and largely so in China today, but it was not the case in China in 2005, when interest rates were fixed by the central bank, with little room for deviation from the so-called benchmark rates. This made it impossible for Chinese banks to accurately price in the risks in the interest rates charged of their customers. Consequently, Chinese banks flocked to large corporate customers—typically state-owned—considered to be of good credit quality. This tendency became even stronger after China's banking reforms in the early 2000s, as bank managers were now held accountable for making bad loans.

SDB's corporate customers were mostly small and mid-sized enterprises (SMEs). In China, SMEs are defined as those firms that have fewer than 1,000 employees and generate less than 400 million yuan (~$60 million by 2021's exchange rate) in revenue in a year. These were typically high-risk borrowers, as opposed to large, conservative, state-owned companies, which were much safer bets. SDB was in no position to compete for that business. Historically the major state-owned enterprises, or SOEs, had strong relationships with the larger national banks. SDB was just too small to matter.

How could SDB grow its corporate loan book if it could not compete with large banks for the best corporate customers, or price in the risks of loans made to SMEs? This was a major conundrum for the bank, which was seeking growth while reducing its risk exposure, two objectives that were inherently contradictory.

Newman was undaunted. He pursued two strategies. One was to find a way to reduce the risks of banking with SMEs; the other was to grow its retail banking business.

Hu Yuefei, now a member of the SPC, had done a good job as manager of the Guangzhou branch, building a quality loan book. While there, he pioneered a new product that he called "supply chain finance." He recognized that while the credit quality of his customers was generally weak, these companies would typically supply goods and services to large firms. He began a program to provide short-term loans to these small suppliers, which were then secured by their receivables—the money they expected to receive from their larger customers. The large corporate buyers were of much stronger credit quality and were unlikely to miss their payments. By relying on the good credit of their buyers, the small suppliers were able to secure much better loan terms. As the receivables came in, the money flowed directly to the bank to pay down the small suppliers' loans.

China's Baidu-Encyclopedia includes an entry under "supply chain finance" that provides a brief history of the product, with due credit given to SDB:

> *In the second half of 2001, Shenzhen Development Bank began to experiment with the business of "lending secured by movable assets and goods" in its Guangzhou branch and Foshan branch [also located in Guangdong Province]. By the end of the year, the total loans booked under this program reached 2 billion yuan…. SDB was the first one in the banking system of China to launch the brand of "Supply Chain Finance" in 2006, and over time, it has lent over 800 billion yuan under the program.*

To me, it was amazing that SDB—such a weak and problematic bank when we began to look to invest in 2002—had one branch manager who had been able to adopt a new product that made such a big difference in the asset quality. Before Newbridge took over, the bank's leadership had never made much of the Guangzhou branch's success. But Newman immediately saw the potential of Hu's idea and launched the product company-wide under the brand Supply Chain Finance. The product enabled SDB to lend money to SMEs without having to take on their risks, and contributed to the rapid growth of SDB's corporate loan book.

In 2006, the first full year under Newman's leadership, the SDB corporate loan book increased by more than 30% from the year prior. Its Supply Chain Finance service won the Second Shenzhen Financial Innovation Award and was voted the best SME finance solution by the China Small Business Association, the China Banking Association, and the *Financial Times* (a Chinese publication, not to be confused with the London-based "pink paper" of the same name).

Meanwhile, Newman made a major effort to develop SDB's retail business. China's retail banking market was still in its infancy. Many consumers were getting their first mortgage loans and first credit cards. With a population of more than one billion and an increasing level of affluence, especially in urban centers, the market potential was massive. After one year, SDB produced a report card that showed great progress in both the quantity and quality of its retail banking loans, which suddenly rivaled the best banks in the world. According to its 2006 annual report:

Retail loans in 2006 recorded an 87.1 percent growth, a rate that put SDB in the number three ranking among national commercial banks. The retail NPL ratio dropped from 2.23 percent at the beginning of the year to 1.24 percent at year end, a considerable achievement given the impact of the bank's historic retail NPL stock. New retail loans in 2006 had an NPL ratio of only 0.07 percent. Valid credit cards rose 49 percent compared with a year ago, and a 71 percent growth was recorded in credit card commission income.

By the end of 2006, SDB seemed to be on a tear. Its total assets had grown by 17% from the year before and its net profit increased by an astonishing 354%, from 311 million yuan ($38 million) to 1,412 million yuan ($177 million) in one year. Newman was delighted with his and SDB's achievements. "It's the fastest turnaround I've been involved in," he exulted to *Institutional Investor*.

We were thrilled too. Newman had proved to be the right CEO at the right time for SDB. He was exceeding all our expectations. Once he settled into his job, he never talked about being an interim CEO or about leaving. He was presiding over the transformation of SDB. He earned the respect of staff members as well as the board. His style of leadership was both hands-on and inclusive; he set targets in consultation with senior officers and with the board. He would listen attentively, gazing directly into the eyes of the speaker. If he agreed, he would say, "That's an excellent idea ..." If he did not agree, he would not brush aside the comment; instead, he might say, "Well, that's a good point, but ..." before making his own point. I had never met a leader as patient and diplomatic as he was.

To take just one example: In the first couple of years of Newman's leadership, we had an independent director named Yuan who was a

holdover from before the Newbridge investment and who, for some reason, was always irritated and antagonistic. At board meetings, he spoke
loudly, almost as if he were in a shouting match. He often insulted Newman directly when he disagreed with him. All of us were appalled by his
antics. I was quite ashamed that a fellow countryman—as he spoke in
an accent of the province my parents had hailed from—was so uncouth.
I would pull him aside during recesses to gently advise him to dial
down. "You would be more convincing without raising your voice," I
pleaded with him.

However, Newman himself never appeared ruffled by the outbursts.
He would wait patiently until the man ended his tirade, and then he might
say, "Well, thank you for your point," before moving on to a point of his
own, or asking others for their views. I was so impressed with how deftly
he defused tensions, time and again. Sometimes, I wondered if his translator had censored Mr. Yuan's rude and insulting words or replaced them
with something complimentary. By and large, the board gave Newman
strong support for his initiatives and plans. And he executed these so bril-
liantly that he never failed to overdeliver.

The market and the financial press were equally taken by the first
foreign chairman and CEO of a Chinese bank. His good reputation soon
spread from within the bank to the market. In 2006, China Central TV,
which had hundreds of millions of viewers, named him one of its 20
"Economic Personalities of the Year." It was an exceptionally rare honor
for a foreigner.

★ ★ ★

These were happy and heady times for us, and for the new SDB. We
were not completely out of the woods, however. For all these encouraging signs and achievements, the bank had a long way to go to truly
rebuild. At the end of 2006, SDB remained weak in many ways; its
capital adequacy ratio was a pathetic 3.7%, and its NPL ratio a too-high
8%. Sure, our capital ratio had improved drastically since 2004, but it
remained well below regulatory requirements. The "bank repairman"
had done wonders, but SDB was still in urgent need of replenishing its
capital. And soon our efforts to raise capital would be stymied by forces
beyond our control.

Frank Newman, chairman of SDB, speaking in 2007 on the occasion of the 20th anniversary of the founding of SDB.

Frank Newman visiting employees on Chinese New Year in February 2010.

Frank Newman signing a Memorandum of Understanding for global supply chain financing with the representative of UPS Capital Corporation.

Wang Bomin, CFO of SDB, receiving a finance delegation from Taiwan in May 2007.

Chapter 16

Game of Chicken

I was formally nominated to the board of directors of SDB in May
2005. The thinking was that any bad feeling that had been left over
from our contentious takeover of the bank had dissipated as the
reforms we'd led gathered speed. It was another few months before
my directorship was approved by the shareholders and then the regulators, although I had been intensely involved in the running of the
bank without any title. Once Frank Newman assumed his position as
CEO, there was no need for me to be so engaged. The systems I had
helped set up, including the Strategy and Policy Committee, continued under Newman's leadership. Now that the bank was in his safe
and experienced hands, I shifted my attention to help SDB raise fresh
capital. New infusions were sorely needed.

The plan had been for SDB to raise capital through a rights issue, which
could only be launched after approval by the securities regulator, the
China Securities Regulatory Commission (CSRC). But conflict arose
between the CSRC and the banking regulator, the China Banking
Regulatory Commission (CBRC). The CBRC, the guardian of China's banking system, was keen for SDB to raise fresh capital, and had
in fact been pressuring us to do so. The CSRC, however, saw supporting the stock market as one of its key (if unstated) responsibilities. In its
view, one way of doing so was to restrict the supply of new stock, and

so it was reluctant to approve capital-raising plans of listed companies, or new listings.

Repeated efforts by the bank staff to obtain the CSRC's approval proved futile. It never said no, but it would not say yes, either. Our application was in limbo.

I went to lobby the regulator at the highest level. In mid-May I took Liu Baorui, the SDB deputy president, and Wang Bomin, the bank's new CFO, to meet the CSRC chairman, Shang Fulin. Shang was courteous but direct. He said it would be impossible for SDB to be given the permission to do a rights issue any time soon. He told us that this had nothing to do with any problems within SDB; it was because all capital raising by listed companies had been suspended. The CSRC had ordered the suspension, pending a "share reform policy" in the works. The door to raising capital in the public markets was shut for the foreseeable future. The meeting lasted less than half an hour.

"Share reforms" would prove to be a seminal event in the development of China's stock market, one that the entire country would be caught up in over the next 12 months.

When China first opened stock exchanges in the early 1990s, many state-owned firms were restructured into joint stock companies before issuing new stock through public offerings. The newly issued shares were listed and traded on the young stock exchanges in Shanghai and Shenzhen. But under the early system, shares owned by the government were not offered to the public. These shares were referred to as "legal person" (LP) shares, and it was this class of shares that had originally made possible our control of SDB. They could not be bought or sold on an actual stock exchange; they could only be traded privately, outside of the stock market. LP shares were thus also referred to as "non-tradable" shares. Under the law, LP shares and tradable shares of the same company were supposed to offer shareholders equal rights, such as the right to vote and receive dividends. Yet LP shares typically were priced at a deep discount, because they were illiquid—they could not be bought or sold in the stock market.

The regulators had long indicated that as China's stock markets matured, this system would need to change to conform with international norms. The trading of LP shares would eventually be permitted on the exchanges. Opening the system in this way was necessary, or the development of China's stock market would remain half baked—in any normal stock market, all shares issued by publicly listed companies were tradable other than those under certain lockup rules for only short periods of time. The trick lay in how to do it without causing too much market turmoil.

Part of the problem was that the volume of existing LP shares was far greater than tradable shares, because typically, new shares issued by a Chinese company in a public offering represented only a minority of its total shares. If the regulators suddenly permitted all LP shares to become tradable, they would open the floodgates for sell orders to surge into the stock market. Already, regulators and the market generally believed that the mere possibility of LP shares being allowed to trade had created an "overhang" effect on tradable shares, thus depressing their value. Now regulators wanted to find a way to convert non-tradable LP shares into tradable shares, removing once and for all the overhang effect, and normalizing China's stock market, but to do so without depressing stock prices. The scheme that the regulators conjured up became known as "share reforms."

The essential feature of share reforms required LP shareholders to compensate tradable shareholders for the right to convert LP shares into tradable shares, on the ground that these tradable shareholders would otherwise suffer losses from the anticipated dip in share prices.

To mitigate the feared depression of stock prices by the surge in the supply of tradable shares in connection with share reforms, regulators also decided to curtail supply by suspending all *new* share issues by listed companies until after they had completed share reforms. Unfortunately, SDB's capital-raising plan had to be suspended indefinitely because there was no telling how long the reforms would take, and when the door to capital-raising would reopen. If we had taken over SDB one year sooner, the rights issue would have been completed before the door was shut. But of course there are no *if*'s in history.

★ ★ ★

I have always thought that in investing, luck, good or bad, is an important factor in the eventual outcome. The other key success factor is acumen or good judgment. But one without the other will not lead to success in life.

While we were unlucky in that share reforms had blocked our rights issue, a stroke of good fortune appeared on another front. GE Capital (GEC) approached us to propose an investment in SDB.

While GEC would ultimately fall on hard times after the 2008 Global Financial Crisis, at the time it was one of the most preeminent financial institutions in the world, with operations in 75 countries. It was expanding aggressively in Asia, and was looking to invest in China's banking sector.

GEC was an investor in Newbridge's private equity funds and regarded SDB as a good opportunity because of our control and stewardship of the bank. While we did not give up on the rights issue, we thought raising some capital through a private placement of new shares with GEC would at least partially meet SDB's capital needs. In addition, we viewed GEC as a friendly party, and a like-minded co-shareholder we could work with.

Chinese banking regulations still required that a qualified foreign investor own no more than 20% of a Chinese bank, and that all qualified foreign investors collectively own a maximum of 25%. As a major financial institution, GEC was certainly qualified. To stay under the 25% limit, we determined that GEC would be able to provide approximately $100 million in capital in exchange for about 7% ownership. That would translate into a valuation at about 2.5 times our investment cost, but still at a significant discount to the stock market price. Newbridge's holdings would be slightly diluted, but the equity capital *per share* would actually increase.

I led the negotiation with GEC, which took more than three months. On September 28, 2005, Frank Newman and Jeffrey Immelt, chairman and CEO of GE, signed the agreement in Beijing.

Although the amount was relatively modest, the GEC investment would bring in badly needed fresh capital for the bank. Both parties were pleased. However, even a private placement by a listed company required CSRC approval. And here our luck ran out. In the next few years, SDB would make repeated efforts to obtain approval for the GEC deal, but it was never granted.

Why did the CSRC refuse? Quite simply, the securities regulator made clear that it would reject a capital-raising plan for any listed company that had not completed the share reforms, this being one of the many types of leverage the regulator could use to force listed companies to undertake share reforms. It felt so frustrating at the time, but it would prove to be a blessing in disguise for us: It meant that we never had to dilute our shareholdings.

Fortunately, the CBRC understood that SDB's failure to raise capital was due to factors beyond the bank's control. They also appreciated the significant improvements we had made to SDB's operations. Therefore, although SDB's capital ratio was still woefully below the regulatory requirements, the CBRC continued its implicit regulatory forbearance, giving the bank time to grow its capital base. And here luck favored us again: China's booming economy had created ample liquidity, enabling the bank to continue to expand its deposit base.

★ ★ ★

The share reforms campaign was by then in full swing, and while we fully understood the importance of fixing China's stock market system, we had no desire to participate. My colleagues at Newbridge and I were perfectly happy to hold on to our non-tradable LP shares. Since we had a controlling stake in the bank, we would likely be able to fetch a premium price by selling that stake to a strategic buyer when we were ready to exit the investment. There was no need for us to turn our shares into the tradable variety, as there was no need to sell them in the market.

It would also have been impractical for SDB to follow the same share reforms as other listed companies. Other firms that had participated in share giveaway schemes had settled on a rate of 30 free shares for every 100 tradeable shares owned—LP shareholders giving their own shares to tradable shareholders. But in most government-controlled listed companies, LP shares were the overwhelming majority—these companies had shares to give away. At SDB they represented only 28% of the total. We couldn't possibly give away that many shares. Newbridge would have had to give away all the shares we held, and we would still have fallen short. SDB initially decided to stay out of share reforms, since it was meant to be voluntary.

We should have known better.

★ ★ ★

The regulators began to subtly put pressure on us to move on share reforms.

According to the rules of the Shenzhen Stock Exchange, the agreement we signed with GEC had to be disclosed in a timely fashion, with an announcement cleared by the exchange. SDB submitted the announcement on the day the agreement was signed. But the stock exchange did not accept it. Officials told us that we had to commit to share reforms before they could clear our announcement. Two days later, the exchange censured SDB for its failure to disclose the agreement with GEC in a timely manner, even though the exchange itself had withheld the announcement. It was like a policeman ordering a driver to pull over, and then issuing a ticket to him for not moving. It was maddening.

In fact, we had always had a good relationship with the stock exchange, and the officials there were helpful in many ways over the years. The exchange could not force us to carry out share reforms, but it wanted to find ways to guide us in what the officials there considered the right direction. They understood that the circumstances of SDB were quite different, and that we could not follow the same share-gifting scheme as other listed firms. But they wanted us to make an effort to get share reforms done one way or another, for their own agenda as well as ours. In my meetings with officials of the exchange, they just smiled when I complained about their discordant demands, the bind they had put us in, and the absurdity of their antics. Behind the smiles there was a gentle reprimand: SDB had to carry out share reforms.

We decided that SDB would come up with its own share-reform plan. It wouldn't be easy. Under the published rules, our plan had to be approved by the securities regulator, and by two thirds of our hundreds of thousands of tradable shareholders. For the latter, it was like asking the question: How many shares of mine should I give to you to make you happy and vote for the scheme? Because of the generous share-gifting plans offered by other listed companies, tradable shareholders had high expectations that we could not possibly meet. We discussed various proposals with the exchange and with large holders of SDB's tradable shares, primarily mutual funds. Despite our efforts, none of our proposals seemed to work for both sides.

Almost one year into the kickoff of share reforms, the Shenzhen exchange was losing patience with us. SDB submitted its audited report to the stock exchange on March 31, 2006, but the officials there informed SDB that the report would not be accepted until the bank had committed to a timetable for share reforms. SDB faced the prospect of breaching the listing rules for failing to publish its annual report in time, regardless of the fact that the regulator was the one preventing the bank from doing so. The consequences would be quite severe—the stock market would suspect something was seriously wrong with our bank to prevent the auditors to issue a clean bill of health on time. We had to step up our efforts and quicken the pace for a share reform scheme.

★ ★ ★

After much work, discussion, and consultation, SDB announced its share reform plan on May 29, 2006. The plan was complicated but the upshot was simple: SDB would pay a special dividend to tradable shareholders only.

The fundamental difference in our plan from others in the market? We had resolved that Newbridge would not give its shares away for free, as they were the property of our investors and we had a fiduciary duty to safeguard them, as a matter of principle. But we agreed to let SDB pay a special dividend to holders of tradable shares if the stock price was traded outside a specified range, up or down. It made no sense for SDB to pay a cash dividend, given its dire need for capital. But we felt compelled to offer this scheme just to get over this hurdle.

The plan was driven by our belief that SDB's stock price was undervalued. Just as the expectation of share reform had created an overhang in the market, depressing stock prices, SDB's inability to raise capital had created an overhang of its own. We believed the stock price would rise if the plan was accepted, for it would mean SDB would then be free to keep growing the bank. We also believed that tradable shareholders, driven by self-interest, would realize that approving our plan would bring a spike in the stock price that would make them a considerable amount of money—even if the value of our special dividend plan paled in comparison with what tradable shareholders in other listed peers received. If they rejected the plan, on the other hand, the negative impact on SDB's share price would likely be significant.

For Newbridge's part, the movement in SDB's stock price, in the short term at least, did not concern us; we had no plan to sell any time soon. But tradable shareholders were in this for a trade, and it would not be in their best interest to draw out this tug of war. If they were rational, we reasoned, they would agree to the plan.

We did not underestimate the difficulties, however. "If SDB is a train rolling towards the peak of success in the future," Frank Newman told one of China's national newspapers in June, "the share reform is the hardest slope to climb at the moment."

Newman was right. At that point, it looked to me highly uncertain whether we would be able to get over this hump. In the next month or so, I devoted almost all my time, along with my colleagues at SDB and Newbridge, to visiting with mutual fund managers and other major tradable shareholders to lobby for their votes. Newman would speak with any major shareholder or any reporter to promote our plan. To one fund manager after another, I explained the difficulties we were in, the progress SDB had made, and why getting over the share reform would pave the way for SDB to complete its capital raising and grow rapidly thereafter. I also presented our plan as the best we could do, given that our backs were already against the wall—we had nothing more to give. Shareholders would shoot themselves in the foot if they rejected the plan, I argued, as failure would harm the bank and all its shareholders, particularly tradable ones. Some remained skeptical but many expressed understanding and support. It appeared reasonable to hope that we would get this share reform plan accepted.

On July 17, 2006, SDB held a special shareholders meeting in the auditorium of its headquarters to vote on the plan. Online voting had started four days earlier. The final tally showed two-thirds of tradable shareholders present at the meeting voted for the plan. But the rules stipulated that passage required approval from two thirds of *all* the tradable shareholders, including those casting abstention votes. In other words, abstention was counted as rejection. When combined with online ballots, 38.92% of tradable shareholders had voted for it, 31.18% against, and 23.79% abstained. The plan had failed by a wide margin to meet the two-thirds threshold.

The stock price promptly dropped by 18% after the vote. As predicted, the losers were the tradable shareholders themselves.

Chapter 17

Window of Opportunity

There is a Chinese saying that when a vehicle reaches the foot of the mountain, there must be a path through it. The Japanese carmaker Toyota became a household name in China through a multi-decade advertising campaign by cleverly paraphrasing that saying: There is always a path through a mountain, and there is always a Toyota if there is a path. We did not know, however, where the path was for our share reforms.

About a week after our share reform plan was voted down, I received an invitation for a meeting with the chairman of Ping An Insurance Group (PAIG), the second-largest insurer in China, about the possibility of a joint investment in China Southern Airlines. On July 28, 2006, I drove to Shenzhen, arriving about noon to meet with Peter Ma, PAIG's chairman and CEO, and Louis Cheung, its president and CFO, at the company's headquarters.

Ma was a paragon of Chinese entrepreneurs. Tall and straight-backed, with close-cropped hair and a slight tan, he looked more like a retired soldier than a businessman. He had founded PAIG in the late 1980s and grown the company rapidly. He had brought in Morgan Stanley and Goldman Sachs as early investors, and by 2006, PAIG was a publicly listed company on the Hong Kong Stock Exchange with a market capitalization of about $15 billion.

Ma greeted me warmly. After pleasantries, we began to talk business. I was interested to hear what they had in mind. It soon became clear, however, that the airline was a smokescreen; Ma's main interest was in a possible investment in SDB. PAIG could help us with share reforms, he said. He was agreeable to the idea of Newbridge keeping control after PAIG's investment, and for us to have a put right to sell our shares at an agreed price to PAIG in the future when we decided to exit.

A put right is an option to sell shares at a future date, at a fixed price. The holder can decide whether or not to exercise the option. For example, say you own one million shares of Company A's stock. If a party gives you a put option to sell the shares at $100 each at any time in the next three years, then that party has a legally binding obligation to buy the shares from you at that price before the option expires, regardless of what the market price is when you exercise it. It is an option because you, the shareholder, will not be obligated to sell to the party.

Ma knew our shares were locked up for five years from the time of our investment and that we were not in a position to sell them at that point. He wanted to make sure that when we were free to sell, we would sell to his company.

It turned out that Ma had been interested in SDB for a long time. I had, in fact, heard a rumor that PAIG was looking at the bank the month before we closed our investment, although by then it was too late for them. Ma had expressed interest in SDB several times before this meeting, but we had never taken the overture seriously; we had just acquired control of the bank and were busy fixing it. Now we were searching for a solution for share reforms, and his offer to help intrigued me. We did not get into specifics, but I left the meeting thinking that a deal was not inconceivable. Certainly, an investment by PAIG in SDB could be mutually beneficial, as long as we continued to control the bank.

Banking was a critical component in Peter Ma's strategic vision. His company had already acquired a small bank and renamed it Ping An Bank. What I did not know was that on the very day I met with him, Ma had signed an agreement to acquire 89% of Shenzhen Commercial Bank (SCB)—the bank formerly managed by Wang Ji, our NPL tsar. Ma and his team must have been busy that day.

PAIG now owned two banks, but neither of them had a nationwide license, so they could operate only in Shenzhen. As a national bank, SDB was about three times the size of SCB in terms of total assets.

Two weeks after Ma had floated his idea, Louis Cheung and Richard Jackson of PAIG paid me a follow-up visit.

A native of Hong Kong, Louis Cheung spoke English with a British accent. He had received a doctorate from Cambridge University, and had previously been a partner at McKinsey & Co. before joining PAIG. He had risen to CFO, then COO, and was now president of the group. He wore light-brown rimmed glasses, and his premature gray hair made him look more seasoned than his 42 years.

Fiftyish and outgoing, Richard Jackson was a 20-year veteran of Citibank and had served as the bank's country manager in South Korea. He had been tapped by Ma to lead the banking business within PAIG, and would later assume the presidency of Ping An Bank, which was now enlarged to include the newly acquired SCB. Just as Frank Newman was considered the most *English* American in China, Jackson would eventually earn a reputation in finance circles as the most *American* Englishman for his bluntness.

The significance of PAIG's interest was hard to overstate. It was a credible suitor, a large and respected institution. The immediate relevance was its offer to help resolve the share reform issue we were grappling with.

In my office that morning, the two gentlemen made a proposal for SDB to issue new shares to PAIG, representing 51% of its total. PAIG would also give Newbridge a put right to sell our shares at a preagreed price after the fifth anniversary of our investment, in line with our commitment to the government not to sell before then. There was no discussion on pricing. They just wanted to float the idea, to see if we would be amenable to the conceptual framework.

I thought the key was valuation. We would be interested if the price was right, not just for us but for public shareholders. In addition, I wondered aloud whether their idea of a 51% stake would appear too aggressive to SDB's shareholders, who would likely consider such an issue excessively dilutive (it would double the total number of SDB shares so that earnings per share would halve). I imagined shareholders might be more receptive to a new issue of, say, 30%. PAIG could buy another 21% in the public market, which would give a boost to SDB's share price. I was also keen to know how PAIG might help with SDB's share reforms, without which none of what they were proposing could actually happen.

Over the next few weeks, these conversations continued. As we exchanged views, I felt there was a significant gap in our thinking, perhaps one that would prove too wide to bridge. On September 5, Jackson came to our office again, at the behest of Ma. This time he was much more accommodating. He said Peter Ma had a strong wish to do a deal with us. He then offered to give Newbridge a put option for our SDB stake that guaranteed us a 35% IRR once we were ready to sell. (IRR, or *internal rate of return*, is a standard measurement of profitability for private equity investments, roughly equivalent to an annualized and compounded rate of return.)

But as he delved into details, I realized that what he offered was not really an option—he wanted to obligate Newbridge to sell our stake to PAIG at a future date, after our lockup period. In other words, our return on the investment would be capped at 35% IRR. An IRR of 35% over five years would translate into 4.5 times our investment cost—turning our $150 million investment into a guaranteed $675 million. That, I had to admit, was a rather generous offer, under any usual circumstances. Yet I did not think it was attractive enough.

The reason was that the management team and we were doing so much good work in turning around and rebuilding SDB that we expected our returns to be even higher. How much higher? We didn't know. But we knew the bank had a lot more potential once it was on the right track of healthy growth. The Chinese economy had registered double-digit economic growth rates in the past three years—2006 would eventually produce a real GDP growth rate of 12.7%—and the momentum was expected to continue for the foreseeable future. Typically, banks do very well in a strong and growing economy. We did not want to be deprived of any upside over a 35% IRR for our investment.

"We appreciate your interest and your offer to help us resolve the share reform issue," I said. "That's indeed our top priority." However, I explained, we would not be interested in capping our return at 35% IRR. We would only be interested in a real put option, that is, we would have the right to sell at the price he was proposing or not to sell.

Jackson returned to our office two days later. He had not liked our reply. He wanted certainty that PAIG would be able to buy our shares and gain control of SDB when we did decide to exit—his objective would not be achieved if we had the option not to sell. He sweetened PAIG's offer by proposing to raise the guaranteed IRR for Newbridge to 45%, *on the condition* that we agree to sell to PAIG when we were ready to exit. Assuming we would sell at the fifth anniversary of our investment, that 45% IRR would translate into six times our investment cost. That, I thought, would be good enough.

★ ★ ★

I shared the latest proposal with my partners, who agreed with me. We now had an understanding with PAIG—no deal as yet, but at least an understanding. Meanwhile we continued to discuss various share reform plans with the stock exchange. On a visit to the exchange on September 6, I once again suggested a number of alternatives, but the officials there rejected each and every one. It seemed nothing would satisfy them unless we were willing to gift our shares, or "cough up some blood," as they put it.

This was a risk that had never crossed our minds when we were evaluating the SDB investment. It was a nonstarter in my view and left me quite dejected. We adjourned the meeting without knowing what to do next. We could drag our feet, but then SDB would be starved of capital. I felt finding a solution was more urgent than ever. A deal with PAIG now seemed to be our best hope.

I had another meeting with Jackson in PAIG's Shenzhen office at the end of September. He brought along a large team, including PAIG's advisors. I was accompanied by my colleagues Daniel Poon, Elaine Chen, and Tak Chung. Jiang Weibo, a lawyer we had retained from the firm Haiwen, also joined us. The body count alone indicated both parties were making a serious effort to reach a deal.

We spent a good amount of time on the issue of share reforms. After some discussion, a plan emerged. The PAIG team was agreeable, as part of the share issuance to PAIG, for SDB to issue 10 warrants for every 10 tradable shares. A warrant gives the right to buy shares at a specified price.

In this case, each warrant would entitle the holder to purchase 0.2 new shares at a below-market price. A warrant whose exercise price is below the market price of the stock is "in the money"—the difference between the market price and the exercise price is like money in the pocket for the warrant holder. Even an out-of-the-money warrant is valuable because there is a possibility that the stock price will rise so the warrant will eventually be in the money. The warrants would be issued only to tradable shareholders, but not LP shareholders. We thought the warrants would be attractive enough for tradable shareholders to approve such a plan.

We also reached the understanding that the bank would issue a certain percentage of new shares to PAIG at 8 yuan per share, which would represent a premium over the average closing prices for the past 20 trading days. Such an issuance by SDB—selling shares to PAIG—was a critical part of the deal and hugely significant for the bank. It would solve, once and for all, its capital inadequacy problem we had been struggling with from day one.

PAIG would also make a general offer in the open market to purchase SDB shares—up to 26% of the total.

Both these moves were expected to fuel a rise in SDB's stock price, making the warrants more valuable and tradable shareholders happy. The entire package should be appealing to regulators as well.

There remained a major issue. If we agreed *now* to sell our SDB stake to PAIG even though we didn't have to deliver our shares until after our lockup period, it would still amount to a "forward sale," according to Jiang, the Haiwen lawyer. That could be perceived as breaching our commitment not to sell until after the fifth anniversary of our investment. Furthermore, if the arrangement was regarded as a forward sale, we might lose our voting rights immediately. Neither would be acceptable to us.

I suggested some ideas to deal with the issue, but Jackson wouldn't entertain any of them. He said his counsel didn't think Newbridge would lose its voting rights, as long as it had not surrendered the shares. But then, after a discussion between the lawyers—theirs and ours—we came to a consensus view that the risk did exist. Here, we hit a snag again.

It was already the start of the one-week National Day holiday. I proposed that the two parties work through the holiday to resolve the remaining issues, aiming to cement the deal before the markets reopened.

But the PAIG team wouldn't meet with us during the holidays, either because they couldn't change their schedules or because they were discouraged by the impasse. I was concerned that the deal might slip away if people left for the holiday, but the PAIG team seemed unconcerned, or perhaps just not motivated to move the deal forward.

★ ★ ★

Newbridge Capital was investing throughout Asia, so I was often traveling to different countries for this or that deal. Now that the PAIG team had gone away for holidays, I traveled to Hanoi for some potential investment opportunities in Vietnam. I was having lunch with the CEO of a major Vietnamese bank when Peter Ma called on October 4. I briefly excused myself from the table to take his call.

"How should we move forward to conclude the SDB deal?" he asked. Too bad his team wasn't as driven as he was, I thought to myself, or as I was, for that matter. I explained why we were stuck; we couldn't accept a forward sale. But I had some ideas to solve the problem, I said, and I would call him back after my lunch if he was willing to hear them. He was.

We connected a few hours later.

"How about if PAIG just issues us an irrevocable and open offer to buy our shares at the IRR we have agreed, but we don't accept it for now?" I suggested. "If you do that, we'll be okay for the rest of the deal to proceed. That way, there is no forward sale."

"How do we know for sure you will eventually accept our offer?" he asked.

"We can't commit to selling our shares to you, or else it would be deemed a forward sale," I replied. "But we can commit to you that we won't sell to anyone else. We will only have the option to either own it forever or sell it to you before your irrevocable offer expires."

"What if you never sell?" he asked.

"Well, theoretically, we do have the right to own it forever," I told him. "But please think about it. It won't be in our best economic interest to do so. We are a private equity investor. We will be motivated to sell sooner or later after the lockup period to realize our profits. And if we let your offer expire, we will get stuck with SDB shares, as we will have committed not to sell to anyone other than to you."

Ma paused for a few seconds. When he spoke again, he seemed persuaded. He asked me to send a note to Jackson to explain my proposal. I was busy going from one engagement to another in Hanoi that day, but I managed to get a note out to Jackson late that afternoon.

When I reached Jackson the next day, he rejected my proposal, citing his counsel's objection. For some reason I could not get my points across to him. Perhaps he was relying too much on his lawyers. Or perhaps he just didn't trust that eventually we would act in our own best interest. I prodded him to ask his lawyers to explain what the problem was with my proposal. I really felt we couldn't just take the lawyers' "no" for an answer.

I was on my way to Hong Kong, planning to transit to Tokyo on Friday, October 6, when I joined a conference call with Jackson and his advisors. After about 45 minutes, PAIG's lawyers agreed that my proposal would remove the uncertainty over our voting rights. It was better for everyone, they helpfully concluded, because PAIG would want us to remain in control of the bank as well until our exit. It would also help sell the deal to many constituencies, including other shareholders and regulators.

Jackson said they needed to reflect further on the plan before talking with us again. I reminded him that we had 24 hours to agree to all the major terms, so we could call a board meeting to approve the deal and halt trading of SDB stock before the stock market opened on Monday. We were, I stressed, already cutting it too close. If we missed this window, then the transaction prices we had been talking about might not hold in the event the stock's market price moved away from us—if the stock price rose significantly above 8 yuan per share, we might encounter resistance from other shareholders to issue new shares to PAIG at that price. We had to seal and announce the deal before the opening bell of the stock market.

However, I didn't hear back from him until Saturday afternoon. He wanted to change the deal by cutting the IRR for our shares from 45% to 40%. Further, PAIG didn't want to do a general offer to purchase secondary shares in the market anymore. I responded that lowering the IRR would be a deal-breaker, although we wouldn't object if PAIG decided not to tender for shares in the market.

It was already Sunday afternoon when I heard from Jackson again. He requested a conference call for that evening. During the call, he wanted to change the deal again, proposing to cut the issue price for SDB shares from 8 yuan to 7.25 yuan; anything higher, he said, wouldn't work for PAIG.

Under the rules of the securities regulator, private placement could be priced no lower than either 90% of the average closing price of the stock over the preceding 20 trading days, or the last closing price. PAIG's new ask was within the rules, as it was higher than 90% of the 20-day average (but below the last closing price), but this last-minute request was a big surprise, and too substantive a change for me to respond to without consulting my partners and the board. Nonetheless, I told him that I would present his terms to them. I was keenly mindful that we were running out of time; we had to halt trading of SDB stock before the exchange opened the next morning.

The Newbridge partners convened a conference call at 7:30 the next morning, October 9. I reported Jackson's last-minute "retrade" on the stock issue price, which had been presented to us on a take-it-or-leave-it basis. We had to make a decision. My partners were disappointed, but after some discussion all agreed the deal should proceed. We all agreed that the benefits of being able to replenish the bank's capital outweighed the discount the bank had to accept to sell stock to PAIG. We still believed it would be positively received by shareholders and regulators alike, especially given that the proposed share-reform plan was part of the deal. We believed the board would be supportive, but there was now no time to "prewire" board members before halting trading. We decided to halt trading and then hold a board meeting on the same day.

It was about 8 a.m. The stock exchange would open at 9:30 a.m. We were literally racing against the clock. But then, a new issue arose. I suppose I shouldn't have been surprised.

It was customary for parties to accept the governing law of a contract to be that of a neutral jurisdiction. We had earlier proposed that the contract between Newbridge and PAIG be governed by the law of a jurisdiction of Newbridge's choosing. We hadn't received any indication of objection and thought this was settled. To our surprise, Jackson sent me an email proposing changing the governing law to be Chinese. His email was clocked at 8:49 a.m.—just 41 minutes before the market was to open. This was unacceptable to us. I counterproposed UK law, which he would not accept. In turn he counterproposed Hong Kong law, and Hong Kong arbitration, in the event of a dispute. His last proposal came at 9:06 a.m.

I accepted it both because I thought it was good enough a compromise and because we were running out of time. I felt his tactic was brinksmanship, but what puzzled me was that he was the one taking the risk of a collapsed deal—a sudden rise in the stock price after the opening would jeopardize the chance for PAIG to complete the deal.

We immediately instructed Xu Jin, SDB's board secretary, to inform the Shenzhen Stock Exchange to halt trading of SDB stock and call an emergency board meeting for that afternoon. PAIG's advisors on stock market matters had earlier advised us that one phone call to the exchange was sufficient to halt trading of a stock, with no need for a specific explanation. They were the only ones with experience in this regard, and so we had all presumed they knew what they were talking about.

Soon, Xu Jin called me back. The stock exchange said it wasn't possible to halt SDB's stock's trading for the morning session. Such requests had to be made before 9 a.m., after which time pre-market orders would be accepted and registered by the exchange for the opening at 9:30 a.m. Furthermore, the exchange officials advised him a request to halt trading had to be made in writing, including reasons and explanations. Under the rules of the stock exchange, once trading started, it could not be halted until the first session finished at 11:30 a.m.

We had no choice but to scramble to dispatch a written request to the stock exchange to halt trading at midday.

As luck would have it, the market opened with a strong rally, and the SDB stock price jumped, along with the market. I quickly drafted our suspension request and the board secretary, who had known nothing of the deal as yet, sent it to the exchange. But I received a thunderbolt, in the form of an e-mail from Jackson: Peter Ma no longer wanted us to halt trading, because PAIG was withdrawing from the deal.

Ma was upset that we had failed to halt trading in time. He didn't know it wasn't our fault and it was his side who had kept making changes until it was too late. I pleaded with Jackson not to withdraw; we would miss our window of opportunity. I also promised that we would hold the agreed price even though the stock had traded higher. But all to no avail. He said his instructions were firm.

I thought there must have been a lack of communication within PAIG. Jackson spoke no Chinese, and Ma's English was limited; perhaps that had been the problem. Ma later told me that he had been on the call with the chairman of the exchange before trading began, and the chairman had not known anything of our request to halt trading. The chairman probably did not know his staff had already rejected our verbal request. But Ma assumed we had never made the request.

Regardless, at that point we had no choice but to go back to the stock exchange to withdraw our suspension request. And I had to call directors of the board one by one to explain why we had called a board meeting, and now were calling it off, all within a span of just a couple of hours.

A rough morning, to say the least. But it was not over.

At 12:01 p.m., Jackson sent me yet another email to say he had now received instructions to proceed with a midday trading suspension. Why couldn't he just call me? I thought to myself. If I hadn't checked my emails religiously, I could have missed so many of his last-minute messages. He gave instructions as if we were not his counterparty but his subordinates, who were expected be on standby at all times for his orders.

In any case, I supposed by then that Ma had learned from the stock exchange chairman that we had indeed made the request in time, but not in a form acceptable to the exchange. Now it was impossible for us to go back to the stock exchange again; we would draw ire for all our flip-flops. Nor could we call a board meeting, having just called and canceled one. We had to wait until the next day.

When all was said and done, there would be no next day. SDB's stock price shot up 6% to 8.69 yuan at the close of trading. It never looked back, reaching 14.47 yuan by the end of the year. Given this surge, there was no chance SDB's board would approve the deal at the price PAIG had wanted. The window of opportunity had snapped shut.

In deal business, like on a battlefield or in so many other things in life, you really have to strike when the iron is hot, to seize the momentum or the moment, to win. If a fisherman slacks off when reeling in a big fish, the catch may snap the line and get away forever.

Chapter 18

Mariana Trench
to Mount Everest

Despite a substantial improvement in the operations and profitability of the bank under our care and Newman's able leadership, SDB remained undercapitalized. Capital was the bottleneck to SDB's growth and transformation. It felt like we had fixed the rudder and restarted the engine, but we could not run our vessel at full steam for lack of fuel. The CSRC still would not approve SDB's capital-raising plans until the bank had completed its share reforms. The proposed deal with PAIG would have solved that problem once and for all, but, by almost a freak accident, that was gone, for good. We were back to dealing with tradable shareholders whose make up was constantly changing, because the stock changed hands every day. But we knew what they wanted: a rising share price, or the prospect of a "gift" from LP shareholders. Without share reforms, neither purpose could be achieved.

A team consisting of senior SDB and Newbridge officers had already spent an incalculable amount of time on this issue. Xiao Suining, who had succeeded Wang Ji as Shenzhen's chief negotiator during our deal, had retired as chief of the Shenzhen branch of the Bank of Communications. We invited him to join SDB as president. He had impressed us as a tough negotiator but also a man of integrity, principle, and deep knowledge in banking, just the type of person we needed. After he came on board, he and I led a team to visit with major tradable shareholders in the hope of finding a plan acceptable to them, and one which we could live with as well.

Xiao was trained as an electrical engineer, with many years of experience in building hydropower plants and other projects. He had a number of patents under his belt, including, most improbably, one for a micromedical device to block blood supply to tumors. He could talk with anyone on a wide range of subjects in-depth, and he carried an aura of authority wherever he went. The bank staff had a running joke about him: "President Xiao knows everything on earth, and half of the things in heaven." He proved most persuasive with tradable shareholders and regulators.

The plan that gradually took shape looked like this: SDB would issue warrants, or call options, that would give shareholders the right, but not the obligation, to purchase new SDB shares at a predetermined price. It was similar to what PAIG had proposed. By issuing such warrants, SDB would be able to kill two birds with one stone: The bank could win the support of shareholders for share reforms while raising needed new capital. Warrants were attractive to shareholders, especially if they were issued in the money. Those who did not want to exercise their warrants could sell them for a gain. The bank would bring in new capital when warrants were exercised. By the end of 2006, SDB's stock price had already risen to 14.47 yuan, roughly doubling the price at which it would have issued shares to PAIG less than three months earlier. SDB would thus be able to raise capital at a much better stock price through warrants.

The only drawback was to the holders of legal person shares, including Newbridge, who would not be receiving any warrants: Our shareholding would be diluted after the warrants were exercised. That would be the price LP shareholders had to pay to make their shares tradable. We decided that this was an acceptable price to pay.

After many meetings with major tradable shareholders and based on the feedback from officials of the securities regulator and the stock exchange, we were prepared for SDB to announce another share reform plan. But now the Shenzhen Stock Exchange would not let us proceed. The official rule was that the acceptance or passage of share reform plans was left entirely to shareholders; there was no need for preapproval by the securities regulator. Indeed, in our first attempt, the stock exchange had not intervened in the process. But this time around, the stock exchange would not let SDB present its plan to shareholders without its approval, insisting that Newbridge and other LP shareholders gift some of our shares or cash to tradable shareholders as part of the arrangement, in addition to warrants.

It was not that the regulators had become particularly meddlesome. They were just fearful our plan would be rejected by tradable shareholders again. The regulators were eager to declare victory for share reforms for the entire market, and they didn't want SDB, one of the last holdouts, to stand in the way. Of course, the more generous LP shareholders were, the higher the likelihood of passage by tradable shareholders. The economic interests of Newbridge were not their concern.

Meanwhile, the securities regulator had issued two *new* rules that would prevent SDB from issuing warrants. The first stipulated that no listed company could issue warrants without having completed share reforms first, which presented a chicken-or-egg dilemma for us. The other required a minimum three-year history of dividend payments for a listed company to be able to issue warrants, a history SDB did not have.

"If you wish to issue warrants, just allow tradable shareholders to buy your own shares at a discount," one official of the stock exchange suggested.

"We might as well give away our shares," I retorted.

Newbridge was in no position to give away shares, as we held them on behalf of our own investors. Our position on that wasn't about to change.

Our first share reform plan had involved the bank paying tradable shareholders a cash dividend. We were prepared to do it again, now in conjunction with issuing the warrants. We thought the securities regulator would like this package well enough to allow us to proceed despite its new rules. But in September 2006, the banking regulator, the CBRC, had issued a new regulation that barred undercapitalized banks from paying cash dividends. This new rule made sense—cash dividends would further deplete a bank's capital—but it was at odds with the wishes of the securities regulator, the CSRC.

We felt boxed in by the new rules and competing regulations. I had to keep reminding myself of a Chinese adage: "Heaven always leaves man with a way out." But where was our way out? I sensed we would have to wait a while for heaven to reveal its mysteries.

For several months, we were engaged in negotiations with the stock exchange without much progress, although all involved professed a mutual interest in helping SDB find a way out of the share-reform conundrum. It should be noted that, officially, the securities regulator was the CSRC, not the stock exchange. But when it came to share reforms, much of the responsibility was delegated by the CSRC to the stock exchange. As time passed, pressure mounted for the regulators to work with us to find a path forward. We were also trying to come up with alternative plans that we hoped would satisfy both the regulators and tradable shareholders.

* * *

Stars finally began to align in April 2007. SDB's stock price had crossed the 20-yuan mark. It would finish the month at almost 25 yuan. The stock was helped by a market rally, but SDB stock outperformed the market, thanks to the bank's strong first-quarter results. Its net profit soared 129% in the first quarter of 2007 over the same period in 2006, and that followed a more than 350% increase in the previous year. Investors now considered SDB a well-managed institution, and the market generally saw significant room for SDB's stock to rise further, if the bank could break its share reform logjam. Compared to the gain in the stock price, whatever "blood" regulators hoped we would cough up in the form of gifted shares would not be very significant.

On April 11, I paid a visit to Zhang Yujun, CEO of the Shenzhen Stock Exchange, and his deputies. I left quite encouraged. He appeared receptive to our plan that the bank would issue a combination of free shares, call warrants, and cash dividends to tradable shareholders only—or at least he did not reject it, as the exchange had done so many times before. Zhang also suggested that we meet and consult with major tradable shareholders to get their feedback, to make sure our plan would receive their support. I thought his advice was sensible.

Over the next few days, Xiao Suining and I visited a number of major tradable shareholders whose combined holding represented about 30% of all SDB tradable shares. This group consisted mostly of domestic mutual funds but also included a number of QFIIs, or qualified foreign institutional investors: major global money managers. Unlike domestic shareholders, the QFIIs generally did not think there was any basis for them to ask for payments from LP shareholders and would support any plan we proposed. They just wanted to get the share reforms done. Domestic shareholders likewise wanted to get over this hump, although they paid more attention to the details of our proposal.

After these visits, I wrote a letter to Zhang, the Shenzhen Stock Exchange CEO, to report the takeaways from our meetings, and to explain how the various shareholders had helped us refine our plan. Their overwhelming desire was to see share reform completed as soon as possible, I reported, so that SDB could raise the needed capital and achieve even faster growth. They all strongly believed that SDB's stock price would sharply rise once the share reform was out of the way, and they, especially the QFIIs, thought their interests would be better served by this than by whatever share gift we could scrape together. They didn't care so much about cash dividends, knowing SDB was already capital-starved. But they all liked the idea of receiving in-the-money warrants. I knew this positive feedback was going to be more effective in swaying the regulators in our favor than if we had simply presented our own case.

These meetings and discussions also had an unintended effect. Sensing that a new share reform program was forthcoming and that it was likely to be passed, the market pundits lit a fire under SDB's stock price. As we were meeting with tradable shareholders, the price rose by the day, which made our job easier; virtually every shareholder we met was now in a good and accommodating mood. I never count on luck in doing anything, but I did feel now that the pendulum of luck was swinging in our direction. By the close of April, the stock was up a staggering 80% from the beginning of the year. The market was cheering us on. The time was right to launch our second plan.

Xiao visited the stock exchange again to discuss our latest thinking. He called me afterward to say that the meeting had gone well; officials at the exchange had liked our proposal and encouraged Xiao to go to Beijing to seek a formal approval from the CSRC.

The new plan had two major elements: SDB would grant stock dividends to tradable shareholders only, and grant warrants to all shareholders, including Newbridge. We were forced to comply with two conflicting rules: The securities regulator would not allow SDB to issue warrants without a three-year history of dividend payments, and the banking regulator would not allow an undercapitalized bank to pay cash dividends. How to reconcile them?

We found one possible way: While SDB could not pay cash dividends, it was allowed to pay stock dividends. SDB would issue to tradable shareholders one free share for every 10 they held. This would dilute the holdings of Newbridge and other LP shareholders by about 7%, which was something of a compromise: We had promised to not gift away the shares we held for our investors, and while we continued to honor that commitment, we were willing to accept the dilution to get the deal done.

SDB's stock price closed at 27.85 yuan per share on May 11, the day the board approved the new plan—a fourfold rise in less than a year. It was not lost on the market that given that day's stock price, the free shares tradable shareholders would receive amounted to more than one billion yuan, a value 50 times greater than the total amount of conditional cash dividends proposed for the first and failed plan.

The second part of the plan involved issuing one warrant for every 10 shares held to all shareholders; but since tradable shareholders would receive stock dividends and therefore more shares, they would also proportionally receive more warrants. The exercise price would be set at 85% of the 20-day average stock price prior to the announcement of the plan.

Xiao had cleared the proposed plan with the CSRC in Beijing on May 8 2007, three days before SDB's board's formal approval. On May 14, the plan was filed with the Shenzhen Stock Exchange, which accepted it on May 16, and the announcement hit the press.

Although all signs were positive, it was hard to tell whether we would be able to garner support from more than two-thirds of all tradable shareholders. That remained a tall order. We decided to lobby them intensively, and to do so in two ways.

The first involved a five-day roadshow. Xiao and I would lead a team to visit with major tradable shareholders, some for the first time and others we had met before, to formally ask for their votes. The other was to engage a public relations firm to devise a media strategy to win over the hundreds of thousands of small shareholders (SDB's 2007 annual report recorded more than a quarter of a million shareholders). I personally worked with the PR firm and met with reporters almost every day to present our case.

Meanwhile, the PR firm did a good job reaching other reporters, providing them with background information and making our case. Reporters were happy that their requests for interviews were never ignored, and often I entertained those requests as the chief representative of Newbridge. The PR firm provided us with feedback from the media, including from online chat rooms, to enable us to respond on a real-time basis. We were thorough. And we received a clear indication that we were winning the PR war when a negative article questioning the performance of SDB was immediately rebuked by a number of rival publications—publications whose reporters had closely followed the SDB case.

★ ★ ★

One of the most important stops on our roadshow was the China Pacific Insurance Company (CPIC), one of the largest insurers in China. Its investment department was apparently quite bullish on SDB's stock. In just a few months, it had acquired enough of it to become the largest single holder of SDB's tradable shares. I met with Yang Wenbin, CPIC's chief investment officer, and his colleagues in their Shanghai office on May 21. I didn't have to explain our plan; they were already familiar with it. He was impressed with the turnaround of SDB and optimistic about its growth potential. There was no doubt he would vote for the new plan. I was quite pleased with what I heard.

Then he switched gears. "We have a proposal to make you," he said. And, after a pause: "We would like to purchase one billion yuan worth of new shares from SDB."

His proposal took me completely by surprise. I had come to talk about share reforms, but I certainly wouldn't pass up any opportunity to raise capital for the bank. I didn't know how serious he was. "At what price," I asked, "do you offer to buy new shares from SDB?"

"Of course, the market price," he replied. "We would offer to buy SDB's new shares at 28 yuan per share."

SDB's stock price had stood at 27.85 yuan before trading was suspended on May 11 in anticipation of the share reform announcement. Usually shares issued through private placement were priced at a discount. But he was willing to pay the full market price. I thought selling new shares to CPIC at this price would be good for the bank and would also send a strong signal to the market that the largest tradable shareholder of SDB was not only supportive of the new plan but also willing to put a significant amount of money where its mouth was.

I was confident that my partners would be receptive to this offer. Sure enough, everyone I briefed by phone was pleased with this turn of events. I moved quickly, instructing the Haiwen lawyer we had engaged for the share reform work to draft an agreement. By the close of the next day, both parties had agreed to the draft document. That was lightning speed in the world of deal making. I was prepared to present the proposed transaction to SDB's board, which had previously been scheduled to convene the next day.

Our primary goal for the board meeting was to finalize the share reform plan. I reported the general positive feedback, but suggested that we make the plan even more attractive to shareholders—we should err on the conservative side, just to be absolutely sure of passage. The time had come to be decisive and to overwhelm the odds. The board agreed. The change I suggested was to increase the amount of warrants, from one to one and a half for every 10 shares. Since LP shareholders would receive warrants pro rata as tradable shares, this would not further dilute our holding. It was a better arrangement for SDB because the bank would receive more capital when the warrants were exercised. We also made adjustments to the valid periods of the warrants. Two thirds would now be exercisable within six months at a price of 19 yuan per share, representing an 18.8% discount to the 20-day average trading price before the suspension; one third would be exercisable within 12 months at the same price.

Unfortunately, the proposed share placement with CPIC had been shelved even before the board was able to consider it. I had asked Xiao to call the securities regulator to get its blessing. The officials there rejected it immediately. The regulators wouldn't consider any capital-raising plan by SDB before the share reform issue was completely out of the way. SDB announced its final share reform plan on May 24. Trading of its stock resumed the next day. The usual rules of the exchange limited daily price movement of a stock to 10% in either direction; but for those companies that had not completed share reforms, the limit was reduced to 5%. SDB hit the daily limit at the opening and stayed there for the rest of the day. We had observed a pattern, not surprisingly, of a strong positive correlation between a company's stock performance upon announcement of a share reform plan and the plan's eventual passage. The jump in SDB's stock was a clear indication that the market liked our final proposal and anticipated smooth passage of it. Internet chat rooms also reflected a change in sentiment. Online, people were expressing a willingness to vote for the plan. We felt there was an excellent chance that the plan would pass this time around.

The special shareholders vote was scheduled for 2:30 p.m. on June 8. Online voting had begun two days earlier, and the yes votes were coming in at more than 98%. I was optimistic. I drove to Shenzhen to join the shareholders meeting. Since the votes by nontradable shareholders didn't matter, my presence was simply to observe this important moment in the history of the bank, and to convey to other shareholders how much we would appreciate their support.

A dispatch by a reporter at *Nanfang Weekend*, entitled "A Behind-the-Scenes Story of Share Reforms," captured the drama of that day quite well. The subtitle of the article was "From Mariana Trench to Mount Everest," which was a phrase he had picked up from a PR consultant we had hired. From the deepest depths of the ocean to the highest peak on earth? In a way, it perfectly described that wild roller coaster ride we had been on for more than five years now. The article described the scene at the SDB shareholders meeting:

Chairman Frank Newman took the just printed data sheet handed to him. He hesitated for a moment and did not immediately announce the results. Instead, he invited all the senior officers of SDB who had sat in the front row to come up to the stage.

After the auditorium had quieted down, Newman announced: the two motions, for the share reform plan and the plan for issuing warrants, passed by high approval rates of 99.05 percent and 96.4 percent, respectively. No sooner did his voice stop before all the senior officers leaped to their feet to hug each other and to shake hands. Waves of applause erupted in the auditorium …

Part IV

Exit

Chapter 19

Coveted Prize

N ow that the news about SDB was positive on several fronts, interest in the bank blossomed.

Among the first to come calling was GE Capital (GEC). It had been more than two years since Frank Newman and Jeffrey Immelt had inked a deal for GEC to make an investment. Then, the securities regulator had blocked it. Now that SDB's share reforms were done, GEC wanted in again.

Under the terms of the earlier agreement, SDB was to issue to GEC new shares priced at 5.247 yuan per share. But now that the stock was trading at about 30 yuan, there was no chance the regulator would approve the placement at such a deep discount. Frustrated, the project leader on the GEC side threatened to sue SDB for breach of contract. The threat was an empty one; the contract clearly stipulated that it was subject to regulatory approval. It was certainly not the fault of SDB that the approval had never been granted.

Given where the stock price stood, in mid-July 2007, GEC offered to revise the contract, raising the issue price to 20 yuan per share. It was highly doubtful the regulator would approve that price, either; it was still about a third below the market. But we felt obligated to GEC to give it a try. We visited the CSRC in Beijing, but once again the regulator was unequivocal: Its rules stipulated that the shares had to be priced at no less than 90% of the 20-day volume-weighted average. GEC was unwilling to pay that price. Despite two years of mutual effort, it was just not meant to happen.

But there was no lack of interest in SDB, and, as we had expected, suitors were beginning to turn their attention to the control stake held by Newbridge. Like PAIG before them, many investors appreciated the growth potential of the bank, especially now that it was on such a strong trajectory. Newbridge remained the only private investor, domestic or foreign, to control a national bank in China. Many coveted such a unique asset.

I had known Yu Yeming at Baosteel, the largest steel company in China, for many years. He was tall and bespectacled with a slightly receding hairline, which gave him a somewhat professorial look. We had worked together back in 1995, when I was still a J.P. Morgan banker, on a deal to hedge Baosteel's Japanese yen-denominated debt. The deal never came to pass, but we had remained friends.

Yu was one of the bright young men in the finance department of Baosteel, a department that would produce many of Baosteel's future leaders. He had recently moved from the top post at Baosteel's finance company to a leadership position, running a number of finance-related subsidiaries of Baosteel Group. Highly profitable and cash-rich, the group had expanded into finance. Baosteel had significant holdings in securities firms, investment trusts, insurance companies, and banks, and ambitions to expand further. Yu was spearheading that effort as head of two Baosteel-controlled companies, Fortune Investment and Fortune Trust.

Yu was thoughtful and full of ideas. But when I sat down with him on an October afternoon in the 40th-floor café of the St. Regis Hotel in Shanghai's Pudong district, I had little inkling of where the conversation would take us. We had spoken before about Newbridge making an investment in Fortune Trust, and about Fortune Investment investing in SDB. But those discussions had been exploratory.

I almost spilled my tea when he brought up a different idea.

Baosteel wanted to acquire our control stake in SDB, at a premium to the market price. At first I didn't think it was possible. Our work to transform and grow the bank wasn't complete, and we thought our control was critical to ensure its health.

Neither did I think that the banking regulator would look favorably at a steel company taking control of a bank. Many countries had laws barring industrial or nonfinancial institutions from controlling banks, such as the Bank Holding Company Act of 1956 in the United States. While China had no such laws, regulators were expected to follow similar principles. Newbridge had also committed to not sell our stake in SDB until after the fifth anniversary of our investment, which was still more than two years away. Since our ownership of a national bank in China was so rare and so difficult to replicate, we were reluctant to let it go, and certainly did not want to exit the investment too soon. Indeed, I sometimes regretted that a private equity firm had to exit its investments to realize value within a few years. I thought an asset like SDB was worth holding for the very long term, both for its rarity and because it seemed the best way to bet on China's continued growth.

But Yu's idea would take care of all those concerns. He proposed that Baosteel acquire all the economic interests of Newbridge's *limited partners* or *investors* in SDB, but still keep Newbridge itself in control.

Typically, the general partner, or GP, of a private equity business makes investment or divestment decisions without consulting its investors, known as limited partners, or LPs. If Newbridge Capital wanted to sell a business, it needed only to inform its LPs afterward, when the transaction closed. What Yu had in mind, however, was for Baosteel to buy out and replace all of Newbridge's LPs in the partnership specifically for the SDB investment, leaving Baosteel the sole LP in the partnership of which Newbridge would remain the GP. Baosteel would continue to compensate Newbridge with the fees and profit sharing that LPs usually pay GPs. For his proposed deal, it would be up to the LPs themselves to decide whether to sell their interests; the GP would have no power to make the sale decision.

Theoretically, the regulatory approvals for such a deal, if required, would not be overly onerous, because there would be no change of control—the same GP, Newbridge, would continue to be in control and in charge. Newbridge would require the consent of its LPs, who I expected would be motivated to sell if the price was right.

I thought his idea was creative and elegant in its simplicity. I could immediately see its merit. I was also amazed that the thought had never crossed my own mind, although it had been staring at me all this time. I thought it could work.

"That's a great idea," I told him. "I have never thought of such a structure. You are really good."

One question I still had was whether what was acceptable to us would still be a good deal for Baosteel, in view of the current stock price. "SDB isn't cheap," I reminded him. "It is traded at about 40 yuan a share."

Yu Yeming had already done his homework. He responded with a question: "Would you take shares of China Construction Bank as payment?"

"Why, yes. Of course."

CCB was one of the four largest banks in China. Baosteel had bought CCB shares at roughly the same time as Bank of America, Temasek (Singapore's sovereign wealth fund), and other strategic investors in connection with the bank's restructuring before its initial public offering in 2005. By my rough estimate, the value of Baosteel's holdings in CCB was now about five or six times its original investment. CCB's market capitalization was approximately $170 billion. As such, its stock was highly liquid—its trading volume in the market was high—and we would have no problem offloading billions of dollars' worth of CCB shares on Hong Kong's stock exchange. I thought CCB shares were not much different from cash.

No wonder Yu was undaunted by SDB's high stock price. It was as though I had bought a Picasso 10 years ago and he had owned a Van Gogh during the same period. Both paintings had greatly appreciated in value and he had already locked in the gains. Now he was proposing a swap. My LPs would get cash after we had sold the CCB shares given to us by Baosteel in exchange for the SDB shares. SDB would be far more attractive to Baosteel than its holding of a tiny minority stake in the CCB, as Newbridge would continue to control it on its behalf.

I was impressed that Yu had already figured out all of this. Still, as always, I had to temper my exuberance.

"Have you received the blessing from your leaders for these ideas?" I asked.

"Not yet," he said. "If you think they work, I will go back to have internal discussions."

I encouraged him to do so.

On October 19, 2007, Yu and I exchanged signature pages by fax for a *Baosteel Group Co., Ltd and Newbridge Cooperative and Investment Framework Agreement*. Baosteel would acquire all the LP interests in Newbridge's investment vehicle for SDB (Newbridge, as the GP, would continue to own a small interest). The price Baosteel would pay was the average of closing prices for the five trading days prior to the date of the agreement; that computed to 40.5 yuan per share. I calculated the total proceeds to be about $2 billion, or about 13 times our original investment cost. It would be a good outcome for our LPs. Newbridge's GP interest would remain invested, and we would continue to operate the bank. The agreement was subject to the approval of Newbridge's LPs.

Meanwhile, we received a bit of good luck courtesy of the calendar. Newbridge was to hold its annual investors' conference in Scottsdale, Arizona, on October 24. The main purpose was for us to report our strategy and investment performance to date. We could now use the occasion to inform them of the agreement with Baosteel. Early that morning, we met with the LP Advisory Committee consisting of representatives of the largest LPs, with the Baosteel purchase topping the agenda. I gave a brief presentation, we had half an hour or so of discussion, and the committee voted unanimously to recommend the Baosteel purchase to all the LPs. Eventually, all LPs gave their consent.

Part of the deal was for SDB to sell new shares to Baosteel. On December 3, SDB announced that it would issue 120 million new shares to Baosteel Group in a private placement. The issuance would represent 5% of SDB's total shares. The issue price would be set at 90% of the average price of five trading days prior to the announcement, or 35.15 yuan per share. The capital raised by SDB would be 4,218 million yuan ($570 million), a solid shot in the arm for the bank.

I was in Beijing on December 28 to visit with CBRC chairman Liu, hoping to gauge his reaction to the Baosteel deal. The weather was bad, the sky dark, and an early morning snow soon turned to sleet and freezing rain. The air was so fouled by burning coal and vehicle exhaust that I felt like I was in a chimney when we were outside. Traffic crawled in the slushy streets.

I arrived at the CBRC office at about 8:20 a.m. and waited for about half an hour before Liu hurried in. He had been stuck in traffic as well. He had only 10 minutes before another appointment. I asked him about the Baosteel deal. He said simply that he did not want industrial firms to control banks. On the proposed issuance by SDB of 5% of its shares to Baosteel, he said he would be fine with Baosteel receiving 5%, but not a board seat. He was in too much of a hurry for a full discussion, but it was clear that he did not want Baosteel to buy out Newbridge's LPs. Even though I told him that Baosteel was not seeking control and that Newbridge would continue to operate the bank, he was quite firm. Then he was off, as hurriedly as he had arrived.

Baosteel continued to make an effort to win approvals but it proved too steep a hill to climb. Soon, there was a change of leadership at the company, and the new leaders were either less keen to pursue the deal or dissuaded by the regulators' opposition. I had a meeting a few months later with Xu Lejiang, Baosteel's new chairman, and Ma Guoqiang, the CFO, in Baosteel's head office. In light of what we thought were insurmountable regulatory obstacles, we mutually agreed not to pursue the deal further.

★ ★ ★

Then, in May 2008, Peter Ma came knocking on our door again.

Ma was a determined man. He employed a group of talented managers for the various businesses of the company and freed himself to spend most of his time thinking, studying, and strategizing for the future of PAIG. He was an engaging interlocutor, sharing and testing his constant stream of ideas with visitors while imploring them to share their own. He was dressed casually and plainly. The only luxury about him was a brand of brown-papered, thin, filter-tipped cigarettes, which, to my nonsmoker's eye, looked special and rare, and which he chain-smoked, lighting up one after another.

Ma was not a man easily deterred by setbacks.

It was clear he wanted to turn his insurance firm into a financial holding company built on three legs: insurance, banking, and asset management. His life insurance business was already the second-largest in the country, catching up fast with the state-controlled China Life. He had created PAIG's banking operation, Ping An Bank, by acquiring and combining two relatively small banks. Still, banking represented less than 10% of PAIG's business; insurance made up virtually all the rest. A financial holding company needed a national bank as its centerpiece; this remained the missing piece in his ambitious plan.

PAIG had narrowly missed its opportunity to acquire control of SDB in 2006. The deal could have been revived if SDB's stock price had eased back to a level closer to the agreed price of 7.25 yuan per share. But it never did. After that date, rather ill-fated for PAIG but incredibly lucky for Newbridge, SDB's stock had climbed continuously and steadily. It reached more than 48 yuan per share on October 31, 2007, more than six times what PAIG would have paid just about one year earlier. In hindsight, PAIG's decision on that day to pull the plug on the deal looked like a colossal mistake.

Coincidentally, around the time PAIG opted out of the SDB deal, another big fish surfaced for PAIG. Guangdong Development Bank (GDB) had been founded in 1988, about the same time as SDB, and was based in nearby Guangzhou. GDB was roughly 40% larger than SDB in total assets, but its reported NPL ratio was about twice as bad as SDB's had been before we took it over. Like SDB, GDB also operated nation-wide. The provincial government restructured the bank, dramatically cut its NPL ratio, and put it up for sale through an auction process. Its national footprint attracted a number of potential domestic and foreign buyers, including PAIG.

Whereas hardly any investors had been interested in a broken Chinese bank when we had signed our framework agreement for SDB, now the sentiment had shifted, no doubt inspired and encouraged by our success. Consequently, the competition for GDB was intense. PAIG emerged as one of three bidders that qualified for the 85% stake. But after an intensive bidding process that lasted more than a year, PAIG lost out to a consortium consisting of Citibank, China Life, CITIC, and China Power Grid. After SDB, this was only the second time that control of a national bank would be sold by the government. It would also end up being the last.

Citibank, which owned 20% of GDB thanks to the same restrictions on foreign ownership that bound Newbridge, was able to secure a contractual right from other members of the consortium to operate the bank. Citi would ultimately sell its stake in GDB to China Life about 10 years later. Had PAIG won, it would have been able to acquire control as a domestic investor. But once again a big fish had slipped through its fingers.

During the first quarter of 2008, China's stock market rally was running out of steam, as global markets showed signs of strain as well. SDB's stock price had also eased, dipping and drifting around 30 yuan. Meanwhile, the transformation and growth of SDB under Newman continued apace. Total assets had increased 35%, with equity capital up 28% and net profit up 88% in 2007, on the back of strong growth in 2006. The bank's reported net profit would have been even greater if it hadn't had to write off legacy bad loans to further bring down its NPL ratio, which now stood at 5.62%, down from the 11.4% figure the bank had reported back in 2004. SDB's capital ratio had risen to 5.77% and was expected to soon exceed the regulatory requirement of 8% when fresh capital came in from the exercise of warrants issued in connection with the share reform. The future of SDB had never looked brighter.

It was against this backdrop that I found myself having tea with Peter Ma on May 8, 2008, at the Shangri-La Hotel in Shanghai's Pudong District. Earlier that day I had spent two hours with Liew Shan-Hock, the head of PAIG's mergers and acquisitions department, to talk about PAIG's renewed interest in SDB.

Liew was a Chinese Malaysian, fluent in English and Chinese. He was one of the talented individuals Ma surrounded himself with, along with the likes of Louis Cheung and Richard Jackson. Ma once told me that he had brought in a team of consumer finance experts, all of whom were Koreans, to start PAIG's consumer finance business. That was unusual in China. Ma would hire and use the best talent in senior positions, regardless of nationality.

Liew told me that PAIG was once again seriously interested in acquiring SDB. I shared with him that there were three things we cared most about, and suggested he discuss them with his superiors.

First and foremost was our reputation. I was referring to how things had gone the last time: PAIG had wanted us to stop, then go forward, and then stop trading of SDB's stock, all in the span of a couple of hours, with no consideration of how these flip-flops might have looked to the stock exchange and our own board.

Second, we would be very particular about *who* would control the bank. We didn't just want to give the reins to the highest bidder. We wanted to be sure that the next controlling shareholder was highly reputable, qualified, and interested in owning the bank for the long term. We cared about the future of SDB after Newbridge. After all, it would be forever tied to our name.

Third, and perhaps most obviously, we wanted to get the best value for our investment.

When I sat down with Ma, he reiterated what Liew had said about PAIG's interest in SDB, adding that he would be keen to work out a deal. I was mindful of the fact that our lockup period—the five-year pledge we had made to hold the bank—would expire at the end of the following year, and therefore it was not too soon for us to explore exit options. In our previous negotiations, we had agreed that PAIG would provide an irrevocable offer for our shares at an IRR of 45% for our investment. SDB had grown significantly since then, and the stock price had quadrupled. Despite the changed circumstances, we would always take seriously a motivated and quality buyer like PAIG. I said we would be happy to explore a deal, although there was no hurry.

The next time I met with Ma and Liew was 18 days later, on May 26. This time it was over lunch at the Shangri-La Hotel in Hong Kong. After some discussion, Ma indicated that PAIG would purchase Newbridge's shares at 39.9 yuan per share, about a 60% premium over that day's market price. The price reflected a control premium for our stake, and was not much below where the stock had been traded before the market correction earlier that year. I thought the offer was fair, and I said as much. A lot of details remained to be worked out, and we agreed the two sides would work together toward a deal.

Chapter 20

Flavor of the Day

The Hong Kong office of Goldman Sachs was in the 68-story Cheung Kong Centre, a slender rectangle of glass and steel that rises above Hong Kong's Central district. The building sits between the sharp-edged Bank of China Tower and the squat gray battleship of the HSBC building. Hong Kong businesspeople are obsessed with *feng shui*, whose masters are said to understand how to harmonize the invisible forces of nature in building designs, channeling their energies to bring about good fortune, or misfortune. It seemed that Hong Kong *feng shui* masters agreed that the I.M. Pei–designed Bank of China Tower, with its vertical, knife-shaped corners, posed a case of bad *feng shui*, particularly for those standing in the way of its energy field. For this reason, the Cheung Kong Center had been designed so that its highest point was just below an imaginary line that connected the tall Bank of China Tower and the short HSBC building. This was a clever effort by the architect to allow the Cheung Kong Center to duck from the negative energy emanating from the Bank of China Tower, and guide it toward HSBC.

Negative energy or not, on June 4, I arrived at Goldman's office at about 9:30 in the morning for a meeting with Peter Ma's three deputies, Louis Cheung, Richard Jackson, and Liew Shan Hock. They were joined by their Goldman advisors. I was alone but prepared. The discussion was friendly and constructive. Both sides were motivated to hash out a deal.

As we negotiated, a storm raged outside. Large raindrops beat the windows so hard that I could see water cascading down the glass like a moving curtain. I savor every moment when I can sit indoors in comfort, insulated from nature's turmoil, remembering the misery I had endured from storms as a hard laborer in my youth in China's Gobi Desert, where there was often no shelter. So it felt good to watch that rain, and I was in good spirits throughout the meeting. It helped that the conversation was going smoothly. By the end, it seemed we had reached agreement on a deal. As if on cue, the rain stopped just as we adjourned, at about 1:30 in the afternoon. We had been together for four hours.

Back in our office, I sent my partners an email under the heading "Sterling—Confidential," summarizing what had transpired. Project Sterling had been our codename for the SDB deal, and we would soon give the deal with PAIG the codename Sterling II. My report reflected my confidence that a deal was imminent and laid out the major terms on which the PAIG side and I had come to an understanding.

First was the deal structure. PAIG would buy out the interests of all Newbridge's limited partners to become the sole limited partner of the partnership that would still be managed by Newbridge as the general partner. The price would be equivalent to 39.9 yuan per share. Newbridge would only cease to be the general partner and pass control of the bank to PAIG after 2009, past the fifth anniversary of Newbridge's investment.

Second, thanks to the share reform plan, Newbridge still owned a bunch of warrants. PAIG would buy them at a price equivalent to the difference between 39.9 yuan per share and the warrant strike price of 19 yuan. The warrants would expire on June 20—only 16 days away. If the PAIG deal did not close before then, PAIG would lend money to Newbridge to exercise the warrants to acquire new shares, which would later be sold to PAIG.

Third, subject to the approval of SDB's board, SDB was to consider issuing new shares to PAIG at 39.9 yuan per share for up to 5% of all SDB shares.

Fourth, subject to the approvals of SDB's board and regulators, PAIG would make a general offer to buy up to 30% of SDB shares in the stock market at a price of 39.9 yuan per share.

The fifth part of the plan was yet to be nailed down with PAIG. After the general offer described in the fourth step was completed, SDB would acquire Ping An Bank, which PAIG already owned, for shares. In an arrangement commonly referred to as a share swap, PAIG would sell Ping An Bank to SDB, not for cash but for SDB shares. After the swap, SDB would own 100% of Ping An Bank; in effect, the two banks would be merged.

Finally, as part of the transaction, PAIG would have some governance rights such as board seats for SDB.

As envisaged, PAIG would become the de facto majority shareholder of SDB after steps one through four were complete. It would end up becoming the direct and indirect owner of 51.68% of SDB, with 30% of shares obtained through a general offer, 5% through a private placement, and a bit more than 16.68% indirectly, by buying out the interests of Newbridge's LPs. But PAIG would become a supermajority shareholder after the fifth part of the transaction, once SDB had acquired Ping An Bank through a share swap. The completed set of transactions would make Peter Ma's dream of having a national bank as the centerpiece of his financial ecosystem finally come true.

I knew that SDB's public shareholders would like the deal, because the agreed price for all the shares, including those to be issued by SDB and those bought in the market, would be 39.9 yuan per share, a 56% premium over that day's closing price of 25.59 yuan. I also knew my partners would like it. One of them, Tim Dattels, a former partner at Goldman Sachs, responded to my email within 12 minutes in his characteristically unreserved and terse fashion: "Pretty amazing, Shan. Fingers crossed."

★ ★ ★

It is so true that the devil lies in the details. And in our case, it seemed that every time we got down to details, all sorts of devils appeared. We at Newbridge had every intention of keeping our commitment not to exit before the fifth anniversary, or the end of 2009. Now it was mid-2008; we still had a year and a half to go. This was why we had proposed a structure similar to the Baosteel arrangement—for PAIG to buy out our LPs, while leaving the GP and GP's interests in place until after 2009. But Liew and Jackson met me again in Goldman's office the next day to say that they didn't think the structure would work. They believed the China Insurance Regulatory Commission (CIRC) would balk at this arrangement.

They were probably right; the CIRC rules prohibited insurance companies from investing in private equity funds or partnerships—so PAIG could not be an LP of Newbridge.

I was also concerned there were too many elements in the deal—PAIG purchasing new shares from SDB, making a general offer for SDB shares held by public shareholders, and selling Ping An Bank to SDB in a share swap, in addition to buying out the interests of our LPs. Each of these elements would have been a major undertaking, subject to regulatory, board, and shareholder approval. It was difficult to tell with certainty if we would be able to obtain all the approvals, and how long it would take for each element in the deal to get done. We could get bogged down if any part of the deal hit a wall or unraveled for one reason or another. I didn't think Newbridge could afford to wait until all the pieces had fallen into place before we were able to complete the sale of our stake. For our own exit, we needed assurances that the deal would get done and a clear sense of timing.

I called Ma on the afternoon of June 9 to share my concerns. After a long conversation, he agreed that the transfer agreement to sell our stake would be unconditional, untied to the other things PAIG wished to do.

Among the other hurdles, preclearing such a deal with regulators would be critical—in this case, the regulatory authorities for banking (CBRC), securities (CSRC), and insurance (CIRC). In addition, both PAIG and Newbridge needed consent from the Shenzhen government, given that both PAIG and SDB were headquartered in Shenzhen. Clearing the deal with the insurance regulator was naturally the responsibility of PAIG. For SDB, the regulator was the CBRC. The CSRC had oversight over both SDB and PAIG as both were publicly listed companies.

★ ★ ★

I flew to Beijing and met with Chairman Liu of the CBRC on June 10. As usual, he was in a hurry. He greeted me warmly but said he had only 20 minutes. Liu was probably the most detail-oriented and hard-working regulator I had ever dealt with. He looked tired. I briefly described the contemplated transaction with PAIG in broad terms but told him we didn't yet have a formal agreement. He thought we were proposing to do the deal after our five-year lockup period, but I told him that the parties wanted to put it into action as soon as possible. He gave no indication of his thinking, asking only that I come back to meet with him and two of his deputies after we had reached agreement with PAIG.

A few days later I met with Chen Yingchun, the vice mayor of Shenzhen, over lunch in the government building's canteen. He listened attentively to my description of the proposed deal, and then said: "It may not be good if you do this before the end of Newbridge's lockup period. That may give people an excuse to attack both Newbridge and the city government."

After all the work we had done to get SDB on track, it was a little disappointing to hear this. But it didn't surprise me. I knew the importance of the five-year lockup for the government. I had thought that this could be solved if Newbridge remained the GP, only allowing PAIG to acquire the LP interests for now. But the government apparently didn't see it the same way.

The next day, June 14, I had a conversation with Fred Hu, head of Goldman's advisory team for PAIG. I had known him for a long time. After earning a doctorate in economics from Harvard, he had begun his banking career as an economist with Goldman. Then he switched to investment banking, probably because he was bored with economic research and wanted to get in on the real action. He was a trusted advisor of Peter Ma.

I reviewed for Hu the feedback from my meetings with the banking regulator and the vice mayor. I told him I believed their concerns could be addressed by Newbridge staying in as the GP, after PAIG had bought the interests of our LPs. And I explained that the PAIG negotiators didn't think the LP structure would work with the CIRC. I asked Hu if he had any ideas as to how we could resolve this conflict.

After some discussion, he suggested we simply receive a put option from PAIG and then support PAIG in doing a general offer. To us, the put option would be better than the structure I had proposed 17 months earlier. That early idea was for PAIG to give us an irrevocable offer for our shares at a specified price. Under that old structure, when Newbridge was ready to exit, it would have been obligated to sell to PAIG, so the option was either to sell to PAIG or to not sell at all. What he was proposing would be a real option for us—either to exercise the option to sell to

PAIG, or not to exercise the option but to sell to other parties at a higher price. I encouraged Hu to discuss the idea with Ma.

I met with Liew Shan Hock, the mergers and acquisitions (M&A) chief at PAIG, and his advisors several times over the next few days. We made good progress in converging on a deal structure. They confirmed their willingness to give us a put option. We agreed to vote in favor of their planned general offer for SDB shares in the open market, as such an offer required the support of SDB's board. There was no reason for us not to be supportive; the offer price would be at a premium to the market price, and therefore attractive to public shareholders.

On June 18 I met once more with Hu and his team. At the end of the meeting, I thought we were close to reaching agreement. After that, I met with Liew. He told me that PAIG now had second thoughts on pricing because of the depressed stock market. In the next few days, Liew informed me that PAIG would be willing to give us a put right, but no longer at the price it had indicated. By then, SDB's stock price had come down about 14% from its close two weeks earlier. Liew suggested that we agree only to a price range, but not a specific price, and the eventual price would depend on where the market stood at the time of the transaction.

I declined. We had been attracted to their offer because we thought their price reflected the true value of SDB. We had no control over the ups and downs of the stock market, and we were not willing to let the market determine the price. We believed strongly in the intrinsic value of SDB, regardless of the market's gyrations. If PAIG wanted to benchmark a deal against the stock market, I told Liew, there would no longer be any basis for further discussion.

Ours was a blunt response, and it must have registered. In early July, Liew took me to lunch at Caprice, a posh French restaurant in the Four Seasons Hotel in Hong Kong. It had been nearly three weeks since our last conversation. I went without any expectations; I just thought it was a good idea to keep the relationship warm on a personal level. But toward the end of our meal, it became clear that he had come to talk about SDB. I was somewhat surprised when he indicated that PAIG would still be amenable to a price of 39.90 yuan a share, and to giving us a put option. But he wanted SDB to acquire Ping An Bank first, followed by an issue of new SDB shares to PAIG. Subsequently, PAIG would make a general offer for

SDB shares in the open market. Just to be sure that I fully understood his proposal, I summarized our conversation in an email later that day.

He responded less than an hour later: "Dear Shan, This conforms with my understanding. Many thanks. I will discuss this internally and with Goldman to check feasibility of the plan. Will revert the soonest."

And then we waited. And waited. But there was no further word.

On August 28, I met with Fred Hu over lunch. What I learned was that some senior staff at PAIG had balked at the idea of giving Newbridge a put option and wanted to obligate us to sell our shares at that price at a future date. I explained that the difference between a put option and a forward sale was that the former would allow us to capture more upside between now and when our lockup expired. We would be free to not exercise the put option and sell our shares to the highest bidder if PAIG would not step up. If, however, we agreed to a fixed price now, we had to receive payment now, but that would be in breach of our lockup obligations. I suggested another idea to solve this dilemma: What if PAIG lent Newbridge an amount of money equivalent to the value of our shares at the agreed price, and took our SDB shares as collateral? Financially, it would be the same as if we had sold the shares, while allowing us to remain a shareholder until after the lockup. I told Hu that either a put option or a loan would work for us.

I was in Shanghai two weeks later when I received a request from Liew for a meeting. He was willing to take the trouble to fly to Shanghai to see me. I agreed, although I didn't know why he was in such a hurry and couldn't wait for me to return to Hong Kong. I figured he must have had some important message to deliver.

The next day, we met for lunch in a Japanese restaurant in Shanghai's Garden Hotel. The hotel was located in the old French Club, a building in the center of what had been the French Concession—one of the many nineteenth-century concessions, or foreign enclaves, that had been carved out of the city before the communist revolution. The French Club was operated by Okura, a Japanese hotel group, but it had retained its classical French décor, including relief sculptures of half-naked figures on the roof staring down at us. The restaurant's Japanese food was supposed to be authentic and first-rate, but neither of us was focused on the cuisine.

After we had sat down, Liew said, somewhat in jest, that Peter Ma was too anxious to get a deal done as soon as possible. "My colleagues and I have to hold him back a bit," he added with a laugh.

That was good, indicating motivation on their side. But I was doubtful whether we could reach a deal now. In fact, I thought that the best time for a deal had already passed. The stock market had been in a rut. SDB's stock now traded at about 17 yuan, down almost exactly one third from its value on June 4, when we had first hammered out the deal in Goldman's office. If PAIG would still pay 39.9 yuan per share, as previously agreed, it would be paying a premium of more than 130% to the market price. That would be great for us, but I doubted PAIG would be able to sell such a lopsided deal to its own board and shareholders.

It turned out that 39.9 wasn't what he had in mind, but what he volunteered still intrigued me.

"Will you not kick me out if I ask you a question?" he cautioned me.

"As long as you pick up the check, I won't," I joked, pointing to the dishes.

"Hear me out first," he said. "Would you consider a price of 35 yuan per share, if we reach agreement on all other things?" he asked.

He thought I might be offended, but in view of the current stock price at about 17, his revised offer was still on the generous side. I didn't give him any indication how I felt but suggested that I would talk with my partners.

In addition to the price, Liew made another offer, no doubt having picked up the idea I had discussed with Fred Hu. He suggested that PAIG provide Newbridge with a loan for the full amount of the purchase price. Upon the expiration of the lockup, Newbridge would simply pay down the loan with SDB shares.

SDB's acquisition of Ping An Bank through a share swap would come next. The ultimate goal was for PAIG to own 67%, or a supermajority of SDB. To get there, it would still need to buy some new shares from SDB and some old shares in the stock market, all at the same price per share as offered to Newbridge.

Liew also made an intriguing offer. Ma was willing to pay Newbridge with PAIG's own shares, or cash, at our option. It was always better to have an option than not to have one.

As I reported in a September 10 email to Bonderman and Carroll, I thought we were getting close to a deal with PAIG, but I remained cautious:

> The caveat is that of course [PAIG] is likely to change its mind on this and that as they have done. But this is the flavor of today. I am scheduled to meet with him tomorrow again if we can agree to the price.

The next day, I had a brief phone conversation with Bonderman. He didn't want to talk about PAIG's new proposal until we saw each other in person as I was soon heading to San Francisco for an internal meeting. Meanwhile, I met with Liew in the afternoon, and we spent another hour together. My impression at the end of the meeting was that if PAIG would not change the price again, we had a deal. The "flavor of today" was a good one.

Chapter 21

Financial Tsunami

O n September 7, 2008, Peter Ma invited me to visit him at PAIG headquarters in the district of Zhangjiang, near Shanghai's Pudong International Airport. Liew Shan-Hock took me there by car, about an hour's drive from the city center. Past the gated entrance, the driveway led through a large and beautiful garden, with lush lawns, tall trees, colorful flowers, and other types of vegetation. There were flowing streams crossed by small bridges, a large Dutch-style watermill, a classical Greek or Roman style pavilion by a pond, and, most improbably, a Gothic-style clock tower some 300 feet high. In the middle of this most un-Chinese scenery stood a sprawling building of seven or eight stories whose ground floor was surrounded by a walkway with arched openings, similar in style to the courtyards of Spanish monasteries.

This palatial setting was home, at least during working hours, to some 6,000 PAIG employees. It had only opened the year before. Ma later told me that he had personally designed the entire layout. He had impressed me as a man of vision and action in terms of his business acumen. In a different way, this place furthered that impression.

Ma gave me a personal tour of the main building. On one floor, there were thousands of workers busy at their computers. Their job was to input data provided by customers. Ma told me that the work process was designed to break up a complicated task into small pieces so that a worker could specialize in one simple aspect. For example, instead of one person putting all the data on a form filled by an insurance customer into the computer, one worker might be tasked with inputting the customer's name, birthday, and ID number, another the customer's educational background, and so forth. That way, the job required minimum training and skill, and could be done efficiently. Once broken down into pieces, an insurance claim for, say, a dental treatment could be processed by three or five high school graduates at a fraction of what it would cost to hire a dentist to do the same job.

Ma took me to the operation control room, which featured a giant electronic screen in front of a manned control panel. The screen showed a bright map of China dotted with red flashing lights that indicated how many insurance tickets PAIG's agents were underwriting at any given moment. He said the entire operational process of his company had been modeled after HSBC, a major shareholder of PAIG, with the help of McKinsey, the management consulting firm, but his own team had greatly improved it.

On the top floor, where the senior management offices were located, large paintings by well-known artists hung on the walls. I was told they were all authentic. Ma's own office was connected to a small indoor garden. We sat in a high-tech–looking boardroom for a brief presentation on his company. Afterward, Ma hosted a lunch, joined by Louis Cheung and Liew, in an Italian-style dining room.

Ma reaffirmed that what I had discussed with Liew in the past two days reflected his own thinking, and that he remained keen to do the SDB deal with Newbridge. He would also welcome a Newbridge investment in PAIG, for which he would be willing to offer us a board seat. I responded positively. I had thought from the beginning that PAIG was a good company for us; what I had seen since then convinced me even further. I had to remind myself, though, not to be too dazzled by what I was shown on this particular day, on Ma's sprawling campus. Even so, I was impressed.

The pace of our discussions quickened after that day. Until now, PAIG had not put any of our verbal agreements in writing. I supposed they wanted to preserve the flexibility to change their mind on various terms. A week later, on the day I was leaving for San Francisco, I received PAIG's proposal by email from Liew. Now I knew they were getting serious.

Their proposed transaction structure involved a few steps, each of which we had agreed to verbally. In effect, PAIG would loan Newbridge an amount of money that would be equal to 35 yuan per share, to be paid back by Newbridge with SDB shares. SDB would acquire Ping An Bank and pay for it with newly issued SDB shares. PAIG would further put capital into SDB by buying new SDB shares.

Liew noted that the offer price for SDB shares represented a 118% premium to the stock's closing price on the previous day, and 85% of its 30-day average. He further noted that the price represented a multiple of 4.9 times SDB's net asset value per share as of June 2008, and a multiple of 20 times the bank's 2008 forecast earnings.

I forwarded PAIG's proposal to Carroll and Bonderman before taking off for San Francisco on the night of September 14.

★ ★ ★

I spent September 15, 2008, in San Francisco for an internal meeting with all my partners, planning to return to Hong Kong late that night. I had little inkling that the day would go down in history as the start of a perfect financial storm, which would soon sweep the globe and devastate the world economy. That morning, Wall Street woke up to the news that Lehman Brothers, the fourth-largest investment bank in the United States, with a 158-year history, had filed for bankruptcy. It was the largest such filing in American history. The news sent shockwaves through global markets.

Throughout that fraught, traumatic day, the Newbridge partners were gathered in the comfort of TPG's sun-drenched San Francisco conference room, quite oblivious to what was going on outside. We had agreed to fully integrate the firm into TPG and rebrand it as TPG Newbridge (eventually the name would be changed to TPG Asia). All of us were there to discuss the restructuring of our business in Hong Kong.

That afternoon, while I was in a car with Bonderman and Carroll, I received a call from a Lehman banker all of us knew well. He told me his job was gone, along with his company. We felt so bad for him that the three of us immediately decided to offer him a one-year consulting contract to tide him and his family over.

The full ramifications of what would later be called the Global Financial Crisis were yet to reveal themselves, but the early signs reminded me of the Asian Financial Crisis of a decade earlier. Over 100 banks of different kinds had failed in South Korea alone during that time. America's banking system today should be much more resilient than the Asian ones a decade ago, I thought to myself. Later events would prove me wrong.

That night I wrote in my journal about the extraordinary events unfolding in the United States and around the world:

> *Today is a day of unprecedented turmoil in the global market. I arrived in SF last night to know that Lehman already filed for bankruptcy, AIG is teetering, and Merrill Lynch has been sold to Bank of America. This has caused the U.S. market to fall 500 points today and the Asia markets including China also fell sharply. Of course, all my friends ... who worked for Lehman are now suddenly out of a job and probably have lost life savings.... It is truly sad.*

While we were together, the Newbridge partners did find time to talk about the PAIG deal. As I had expected, it took no time for us to agree that the deal described in Liew's proposal was attractive and basically acceptable. We had matters to clarify, questions that I raised by phone with Peter Ma and summarized in an email I sent Liew before I departed for Hong Kong that night:

> *We had some good meetings on our mutually interested subject in San Francisco today. We're very focused on this but we still have some work to do internally. I called Peter just now and asked three questions: (1) Is there any room for price improvement? (2) When you say, "subject to due diligence," under what conditions would you want to renegotiate price assuming we agree to a particular price? (3) We assume that we'll both make the best effort for all the steps to happen but the first step (i.e., exchangeable bond [this would be the loan from PAIG to Newbridge which could be exchanged into the SDB shares]) isn't conditional on, or subject to, other steps. Is this assumption correct? (We raise this question because we all know the time for regulatory approvals for*

*these other steps can be much longer). I'll appreciate it very much if
you can help me with these questions before you close today.*

I closed the letter by observing:

*The market looks quite bloody today.... It seems that the [U.S.] gov-
ernment will be forced to do something much more than what it was
willing to over the weekend to prevent a systemic calamity.*

I received a reply on my arrival in Hong Kong. PAIG would not hold
firm on the 35 yuan per share price. What raised further doubts in my
mind about PAIG's commitment was this statement:

*At the next stage of "preliminary due diligence," we reserve the oppor-
tunity to renegotiate price, if we found material discrepancies in respect
of NPL recognition and/or provision adequacy, significant unreserved
other liabilities and other significant issues that may affect value.*

I took his words to mean that their offered price was like a reflection of the moon in water: good to look at but impossible to touch. We
knew that loan classification could be subjective, arbitrary, or subject to
disagreement. If they intended to renegotiate the price, then the basis of a
deal would be quite weak. We had no intention of compromising further
on the price.

He also said that the deal had to "have all the four steps intercon-
ditional," which meant that everything would collapse or be delayed if
PAIG encountered difficulties in obtaining approvals for any of the individual pieces in the proposal. Whereas his written proposal had lifted my
hopes for a deal, I now thought the prospect was rather dim. I suspected
that the dramatic events in the markets over the past few days might have
given them pause. Perhaps their "clarification" was really a way to grace-
fully back out of their own proposal.

Although we didn't like PAIG's conditions, I wanted to sit down
with Liew to make an in-person effort to come to terms. But at this point,
I found it difficult to nail down a meeting, which signaled to me that
either he was distracted by some other fires to put out or that PAIG was
getting cold feet. Either way, the PAIG side went quiet, the momentum

was soon lost, and indeed, the deal appeared to have become a victim of the global financial storm.

And of course the storm only strengthened, and spread. Facing a systemic calamity, the George W. Bush administration took action. While I was in the air between San Francisco and Hong Kong, the U.S. Federal Reserve Board announced it was bailing out AIG, the insurance titan, by injecting $85 billion into the company. The AIG rescue package would balloon to $180 billion in two months. Almost all the major financial institutions, including Citibank, Bank of America, Goldman Sachs, and Morgan Stanley, were teetering on the brink of collapse. The damage spread quickly to the overall economy, as unemployment rates soared in the United States and Europe.

The waves of this financial tsunami soon crashed onto every shore on the globe. Banks in Asia, however, would prove surprisingly resilient. There were almost no reports of bank failures or government bailouts in Asia, where banks and regulators alike had learned a lesson from the 1997 Asian Financial Crisis; banks had built up their capital base and substantially reduced risk exposures. So when the 2008 crisis struck, Asian banks were by and large strong enough to withstand the shock, so they were spared the carnage engulfing their Western counterparts. Nonetheless, the stock markets in Asia took a beating along with the rest of the world.

China's stock market had been in a downdraft for much of 2008. As the markets elsewhere plunged, the downward trend in the Chinese market accelerated. SDB's stock price, which had already lost much ground from its highs in the 40s (yuan per share) in 2007, now slid to the single digits. It closed at 8.37 yuan at the end of October 2008, less than half the September 9 price, when I had last met with Liew in Shanghai. Under the circumstances, I had no illusion that PAIG, or anyone, would be willing to do a deal based on the old prices.

PAIG's stock was faring even worse. A little less than a year earlier, it had spent $3.5 billion for 5% of Fortis, the Belgian-Dutch financial firm. Fortis blew up spectacularly, and was nationalized by the governments of Belgium, the Netherlands, and Luxembourg in September 2008. Soon after, PAIG announced that it would book a loss of $2.3 billion on the deal—two-thirds of its original investment. Its stock price, which had already been declining along with the rest of the market, took a drubbing. Its A-share price tumbled from an all-time high of 149.28 yuan on

October 24, 2007, to a low of 19.9 yuan on October 28, 2008, a loss of 87% in just a year.

What a difference that one year had made. By October 2008, the Dow Jones Industrial Average had come down 42% from its peak a year earlier, and the rout in the United States was just beginning. The Hong Kong, Shanghai, and Shenzhen exchanges all tumbled as well, off a staggering 65–72% in October 2008 from their 2007 highs.

Ironically, SDB's actual operations were stronger than ever. In 2008 its total assets grew another 35% and its operating profits grew 41%. It just showed how the stock market was not always a good yardstick of value.

But when it came to PAIG—or any other suitors—it didn't matter at that point how strong our balance sheet was. All the stars needed to be aligned for a deal to happen, as Bonderman liked to say. Now the stars were just a scattered mess.

It was against this backdrop of financial turmoil that Peter Ma invited me to yet another meeting in his Shanghai office. Over lunch on November 7, with Liew in attendance as well, we talked again about the deal. Ma focused the conversation now on a share swap, in which PAIG would buy our SDB stake with its own shares.

While I had considered it an option for us either to take cash or shares of PAIG in exchange for our SDB shares, now a cash deal was practically impossible. The premium PAIG had to pay to entice us to sell our SDB stake would be too high for it to swallow, given where our stock price stood. In light of how much both stocks had fallen, a share exchange made more sense because it would represent value for value.

Still, I had several concerns that prevented me from embracing the idea. We would be swapping our control position in a bank for a minority position in an insurance company. We could always improve the intrinsic value of the bank under our control, but we would only be taking a ride as a minority shareholder of PAIG. Not a good trade, I reasoned, unless we could sell SDB for a significant control premium. Moreover, when and for how much we would be able to realize the value of our SDB investment in cash remained uncertain; we would be completely at the mercy of the market. Also, Newbridge would have to carry out significant due diligence before deciding whether to take PAIG shares. PAIG was a huge, sprawling business.

Before we parted company, I offered to think further about how to build some other elements into the structure to address these concerns.

On November 14, Bonderman came to Shanghai at my request, in connection with another business under our control there. I took him, along with one of my Newbridge partners, Tim Dattels, to visit Ma and his team. Ma was a gracious host, as ever, taking us on a tour of the corporate campus. By now I knew that tour well. But it was an eye-opener for Bonderman and Dattels. This was followed by a meeting during which Ma laid out in detail his vision and plans for his group. We then moved to the executive dining room, where we enjoyed a good meal while continuing our conversation. We ended up spending two and a half hours there. By the end of the visit, Bonderman had warmed considerably to the idea of a share swap. To me, it seemed to be the only way forward.

We still needed to address a few of my concerns. Whereas Ma wanted a swap of shares based on their respective market values, I thought it only fair for us to get a control premium, which, after all, had been a tenet of previous deals we had considered since 2006. I also thought the downside of holding PAIG's common stock was too high; we needed a PAIG security that would keep a floor to the value of our investment. Also, if the deal required SDB to acquire Ping An Bank for SDB shares, how would we determine the relative value of the two banks? And would the answers prove palatable to the public shareholders of both companies?

I shared my thoughts with Dattels via email on November 23:

> *It seems that David is willing to look further into the swap of securities idea regarding SDB,.... I am supportive of a swap deal, subject to agreements on detailed terms and I also think there is much advantage to get a deal done sooner rather than later because [Ping An] Bank is moving fast on its own and [SDB] needs capital. As we were walking out of [PAIG] the other day, Peter asked me what premium we would want for [SDB]. I asked him to give us a proposal. History: in 2006 October, [SDB] was traded at 7 per share and Peter offered to buy us out at 22 per share.... More recently, the offer was about 40 ... when the market was about 20 or below. So it seems to me that if he proposes a control premium of anywhere above 100 percent to the current market, it is within the fair range. The current price is about 10. If we are talking about 22, we are back to 2006 level. At*

this level, I wouldn't sell for cash as the market is so depressed and SDB is traded at about 7x [price to earnings multiple]. But [PAIG] share price is equally depressed. If we can get a [PAIG] security, it would be okay as we can take a ride as the market rises.

In the same email, I floated another idea to mitigate our risks:

Regarding what security we want from [PAIG], I am thinking of a tradable CB [convertible bond], preferably, or a convertible preferred, both with a reasonable coupon [interest], with conversion price at the current market.... So we have a floor to our exit price and we possibly will get more upside along with [PAIG]....

A convertible bond works in such a way that it allows the holder to convert the bond into shares of the same issuer at a predetermined price. So if the stock price rises above the conversion price, the CB holder may convert the bond into shares to capture the price differential. If the stock is traded at or below the conversion price, the holder would receive the principal value in cash when the bond matures. The instrument allows the investor to get the upside of stock price appreciation without the downside if the stock falls in value.

I also considered other necessary elements of a deal:

The complexity is that as part of the transaction, [PAIG] will need to inject [Ping An] Bank into [SDB] for shares. There, I am thinking of a merger ratio based on net asset value, as the two sides discussed before.... [SDB] will also have to issue new shares to [PAIG] at the current market or slightly above to placate public shareholders, in order to get to the percentage [of shareholding] it needs....

Dattels took this all in, and, at his suggestion, we went to our own Investment Review Committee (IRC) on December 8, to seek an in-principle approval for a deal with PAIG. In our memorandum to the IRC, we proposed selling our SDB stake for a convertible bond issued by PAIG. Separately, we proposed investing up to $1.5 billion in PAIG. The memo laid out the various steps necessary to allow PAIG to ultimately become a 51–60% shareholder of the combined entity of SDB and Ping

An Bank. If we became a shareholder of PAIG, what did we see in it? The memo explained:

> In the mind of [PAIG], SDB fills the missing link or gap in [PAIG]'s long-term strategy. Peter Ma wishes for [PAIG] to build on three major pillars, insurance which includes both life and P&C [property and casualty], wealth management and banking.... [PAIG] sees great synergy in cross selling opportunities for the three businesses with an integrated back office....

> ... Peter wishes to turn the entire network of PAIG into, effectively, an expanded branch network for the combined bank and his 40 million plus customers into bank customers. The ability of offering multiple financial products will enable [PAIG] to expand its customer base from 40 million currently to 100 million in five years, as envisioned by [PAIG]. For that purpose, he also wants to create an opportunity for us to make a significant investment in [PAIG] as he appreciates our knowledge in global investment opportunities, in China and in banking. He likes the idea of making [Newbridge] a partner and he believes HSBC [a major shareholder of PAIG] will be supportive or at least agreeable.

The IRC unanimously supported further negotiation with PAIG.

On December 11, a Newbridge team, comprised of Carroll, Dattels, Jessica Li (a new member of the team), and myself, visited PAIG's Zhangjiang campus. We met with Ma, Cheung, Jackson, and Liew for lunch, followed by further discussion. We stayed for six hours. During the meetings, I proposed that PAIG price our SDB shares at a 100% premium to the market. On this basis we would exchange our SDB shares for convertible bonds issued by PAIG, with a conversion price around PAIG stock's then–market value. We had a good discussion on the deal structure and its various pieces. But Ma balked at the idea of having to price our SDB shares at a premium.

★ ★ ★

At this point, SDB's board made an extraordinary and bold decision, based on a proposal by Frank Newman. He suggested that SDB write off almost all the legacy NPLs in one go. Even without the proposed write-offs, SDB's management believed the bank's classification of its loan portfolio was prudent and its provisions for NPLs sufficient. Nevertheless, Newman wanted to take advantage of SDB's strong profitability in 2008 to clean up its balance sheet once and for all, in anticipation of a tougher atmosphere in 2009 and beyond.

What made Newman's decision particularly bold and extraordinary was that the capital required for doing this amounted to more than $2 billion, a huge sum for SDB.

The NPL write-off in Newman's proposal alone came to a whopping 8.6 billion yuan ($1.3 billion). He was also proposing an additional provision of 5.5 billion yuan ($810 million). It was unusual to write off substandard loans, which typically required about 20% provision. But Newman felt this was the right balance to strike to reach a provision-to-NPL coverage ratio of more than 100%. He believed this would enable SDB to stand out as the cleanest bank in China. Eventually, the provision was increased to more than 7 billion yuan ($1.1 billion), an amount that would wipe out all but about 600 million yuan of net profit for 2008. It was an especially audacious move given the expected negative impact on the stock, and the fact that we were in the midst of talks on the sale of our stake to PAIG.

Under Newman's stewardship, SDB had steadily brought down its NPL ratio from a nominal 11.4% in 2004, which we all knew actually understated the size of the bank's problem loans, to an actual level of just 5.6% in 2007. The new loans booked since our takeover had proved to be of excellent quality, with a negligible nonperforming ratio, thanks to SDB's new, tighter risk-control system.

In 2008, the banking regulator issued a new rule requiring banks to gradually increase their NPL coverage ratios (i.e., the ratio of the money set aside for the NPLs to the amount of NPLs) to be no less than 150%. If a bank had 100 yuan of bad loans, it had to set aside 150 yuan or more for it. Consequently, it would be much more costly to keep an NPL than to write it off.

I had asked the banking regulator Liu Mingkang why he imposed tough NPL coverage ratios. He said he did not trust the NPL levels reported by Chinese banks and believed that a large amount of loans classified as "special mention" would eventually migrate into "substandard"

or NPLs. A tougher policy would force banks to write off NPLs to maintain a clean balance sheet, which would in turn make the banking system
healthier. That, I thought, was the best policy move on the part of the
banking regulator to squeeze NPLs out of banks.

Although the new policy provided the impetus, SDB had a choice.
It could have decided not to write off all its NPLs, in order to report far
better profitability. But Newman decided it was time for the bank to rid
itself of almost all the NPLs now that he had the means to do so: In the
last year alone, the profit before provision and taxes had risen 41% to more
than 8 billion yuan ($1.17 billion by the exchange rate at the end of
2008). We supported the move because we believed any smart strategic
investor would appreciate and care more about the long-term potential
of the bank than its short-term profitability.

Newman later recalled his decision in a private email to me on
November 1, 2021:

> *I did not have any doubts about the write-offs.*
>
> *The CBRC had set the stage in late 2008. It said that coverage
> ratio requirements for NPLs were going up and up, heading towards
> 200 percent or more. Eventually, the industry standard did move up to
> more than 200 percent, depending on the severity of the classifications.
> And it encouraged "clean-up" of NPL portfolios before year-end.*
>
> *The coverage ratio looked like a significant growing burden on the
> bank's future.*
>
> *The combination of the two seemed like a golden opportunity.*
>
> *SDB had about 12.5 billion [yuan] NPLs at year-end 2007, with a
> Loan Loss Reserve of about 6 billion, and an annual provision (addi-
> tion to the Loan Loss Reserve) of about 2 billion for 2007.*
>
> *But 12.5 billion times 200 percent was 25 billion yuan! The growing
> coverage ratio requirement could mean a need for another 15–20 billion
> yuan added to reserves, from earnings over the next couple of years.*
>
> *The arithmetic seemed very clear: a 200 percent NPL coverage meant
> a charge to earnings over time of 2 yuan for every one yuan of NPL.*

That meant that if we could write off 10 billion of NPLs, the bank could save 10 billion in earnings! That's what we did.

We wrote off about 10 billion, leaving less than 2 billion NPLs—all of the lowest risk category—at year-end 2008. At year-end 2008, a 150 percent to 200 percent coverage ratio would amount to about 3–4 billion, and we already had 2 billion left in the reserve.

There was the other, intangible, benefit, too. The high NPL ratio hung like a black cloud over SDB, in the minds of the CBRC, investors, and the public—"a bank with significant credit problems." The opportunity to take the bank from the dark cloud to bright sunshine, as one of the very best, cleanest banks anywhere, offered big benefits in addition to the very tangible benefits to earnings.

I was very confident, based on the arithmetic and stepping out from under the dark cloud.

In the end, he would prove absolutely right: Earnings and capital rose very strongly in 2009, the capital ratio reached its highest level in many years, and the bank stood out as the cleanest major bank in the country.

I met with Richard Jackson and Liew Shan-Hock of PAIG on December 19 in Hong Kong to brief them of our decision on the write-off and the implications for SDB's annual financial results. I knew that Jackson, as a veteran banker, would appreciate our plan to clean out the NPLs. But I could also see that they were surprised and anxious about the magnitude of SDB's plan. They wondered how much it would depress SDB's stock price, which of course would in turn complicate the deal between us.

I wasn't worried. We were making the right decision for the long-term health and growth of the bank. Besides, I sensed PAIG was not yet prepared to move forward with our deal any time soon; we were still far apart on the premium we wanted for our shares. PAIG had repeatedly insinuated its intention to renegotiate the price after conducting due diligence on SDB. I thought the subtext of their message was that our relatively high NPL ratio would give them ample reason to cut the price further. Now, I thought, no more excuses. Of course, if SDB's stock price were to tank, it would make it difficult for the parties to reach a deal.

Indeed, SDB's massive write-off plan had spooked them, as I soon found out in a meeting with Ma in his Shenzhen office on December 23. He said PAIG's capital market staff were concerned that our large-scale provisioning and write-off would depress the SDB stock price for some time. He had indicated earlier that he would consider a premium of 60–100% for our SDB stake in a share swap. Now he was not even firm on 60%. Instead, he proposed that PAIG conduct in-depth due diligence on SDB before talking about price.

I didn't think that would work. Both SDB and PAIG were public companies. SDB had to make a public announcement for such an exercise. If we did so without having at least agreed to a price range, the risk of no deal in the end would be too great. It would be damaging to our bank and its reputation if PAIG were perceived as having kicked the tires and then walking away. Nor would we want to conduct due diligence on PAIG without reasonable assurance of a deal, as the process would be costly and time-consuming.

Through strong earnings and capital injection when the 2007 warrants were exercised, SDB's capital adequacy ratio had reached 8.6% in 2008. But now the CBRC had raised the bar for all Chinese banks, mandating a capital ratio above 10%, from 8% earlier. It asked that SDB raise more capital through a private placement. Newbridge was prepared to invest more in SDB, but we could only increase our shareholding to 20%—the foreign ownership limit. If we had had a deal with PAIG, we would have invited PAIG to participate in the private placement. But now that the talks with PAIG were put on the backburner, that option was off the table.

I left Ma's office after two and a half hours on that day before Christmas Eve, and returned to Hong Kong. As my car reached Central, I saw all the buildings that were bright with cheerful and colorful Christmas lights. The air was filled with Christmas melodies. But my hopes for a deal with PAIG had dimmed. At least, as I reported to my partners in an email that evening, "It doesn't seem that there will be a deal with PAIG any time soon."

Chapter 22

Stars Align

Christmas 2008 brought little holiday cheer to the world. Too many people—investors, consumers, employees, and their families—had been punished by the crushing waves of the global economic crisis. A number of major financial institutions in the United States and Europe had failed. Personal bankruptcies soared, as did unemployment rates. Western banks were still in serious trouble, despite the unprecedented bailouts and infusions of capital by the U.S. and European governments, which were doling out what soon amounted to trillions of dollars.

Away from the epicenter, Asia continued to weather things a little better. China registered an economic growth rate of 9% for 2008—impressive for sure, although it was the country's lowest figure in seven years and the annualized rate for the fourth quarter was only 6.8%. By its own standard, China was in its worst economic slowdown in a decade. The Chinese stock markets were trading in bear territory. Nobody knew how bad things could get, how and when the crisis would end, and how much more damage might be done before then. Investors and policy-makers alike were trying to sort out what had gone wrong, and what could be done to pull economies out from the malaise.

Over the holidays I traveled to Whistler, the ski resort in western Canada, with my daughter LeeAnn for a much-needed vacation. I did my best to escape a bit, although I will confess that there were moments when I was checking my phone from the chairlifts. I also had some time to reflect on all these dramatic events, and I wrote a couple of opinion pieces to collect and share my own thoughts and observations. One of these was published in the *Wall Street Journal*, with the title "The Seoul Solution to the Banking Crisis." The title, given by the paper's editor, was a clever pun referring to what Newbridge had done in the restructuring of Korea's banking system during the Asian Financial Crisis a decade earlier, and why I considered it to be the sole or only solution to the current crisis.

In the article, I explained that in our acquisition of the failed Korea First Bank, we had created a method that allowed both the government and investors to get a better deal from the toxic assets than a simple negotiated sale-and-purchase arrangement. I wrote that the same method could be applied to resolve and liquidate the toxic assets the U.S. government was now taking off the hands of banks. The opening paragraph addressed both the disastrous state of the American financial system and the dilemma the government now faced:

> *Let's face it: the American financial system is basically insolvent. To date, the U.S. government has committed, on behalf of taxpayers, more than $7 trillion of capital injections and guarantees to financial institutions. Treasury Secretary Timothy Geithner said Tuesday the government will pour up to $1 trillion more into a "Public-Private Investment Fund," which will be tasked to buy up banks' bad assets— the real blockage in the credit pipeline. The trouble isn't, however, that banks don't want to sell loans. They just don't know what a fair price is in a now-illiquid loan market because there are no buyers.*

How to solve this problem? I explained what we had done in Korea a decade earlier:

> *Since the market was illiquid, we realized that it was impossible to determine the "fair value" in the near term. We thus agreed to a so-called future "buy or sell" arrangement. Over the following three years, on the anniversary of our agreement, the bank would name the price for any existing loan on its books, and the government would have the option to "buy" or "sell" that loan at that price.*

The goal was for the government to minimize the amount of money it would have to inject to make up the difference between the market and face values of bad loans, and for us not to have to bear the losses from the bad loans we had inherited when we bought the failed bank from the government. This arrangement gave us the time to work out or to improve the value of these loans, and perhaps for the loan value to recover over time.

I closed the piece by suggesting that the U.S. government should use the same methodology to sell toxic assets to private investors, clean up the troubled banks it had rescued, and maximize recovery of value from the assets. This method would save taxpayers money and incentivize private investors to help clean up the battered balance sheets of those banks.

★ ★ ★

When my flight from Shanghai landed in Hong Kong on the afternoon of January 7, 2009, I found an email from Peter Ma requesting a call. I had expected him to reach out; even though we were far apart on the pricing of our SDB stake, I knew he wouldn't simply abandon the idea of a deal, at least not without further conversation. He told me he had spoken with Chen Yingchun, the vice mayor of Shenzhen. He said Chen was supportive of the deal.

"I haven't spoken with my team as yet," he said, "but I wonder if we can both make an effort to come to a compromise?"

"What do you have in mind?" I asked.

"We won't be able to accept your ask of 100% premium. How about an 80% premium over today's market price, or 18 yuan per share?"

I actually had not checked the closing price. When I did, I found that it had settled at 9.9 yuan, down from 10.30 the previous day.

I did not directly answer his question but said that since we were talking about an exchange of securities, we needed to fix the conversion price for the convertible bond (CB) I had proposed that PAIG would issue in exchange for our SDB shares. If the price at which we could convert the CBs into PAIG stock also carried a premium over the market, that would offset whatever premium we agreed on for SDB's stock. That would not be acceptable to us.

Conversely, "if the conversion price for your CBs represents a discount to the market price of PAIG stock, then we can bridge the gap," I suggested. "We need to fix the two prices at the same time, which means we need to fix the exchange ratio between the two stocks."

I tried to draw a parallel. In September 2008, Warren Buffett had invested $5 billion in Goldman Sachs for a preferred stock carrying a 10% annual dividend, and a warrant that allowed him to convert the investment into common stock at an 8% discount to the closing price of Goldman's stock the day before the announcement. What Buffett received was in effect a convertible preferred stock, similar to a convertible bond, except that preferred stocks did not have a maturity date, whereas a CB does. The point I wanted to get across to Ma was that the conversion price for Buffett's deal had been a discount to the market price. And that, I thought, served as a good benchmark for what I was asking. At any rate, I figured citing Buffett couldn't hurt.

Ma seemed to have anticipated my question. "My team," he said, "wants the conversion price to be some kind of an average over a period of time."

I thought that might work. I knew PAIG's stock price had been rising from its lows of the previous year, so that any kind of average, whether one-month or two-month, would represent a discount to the current market price. If Ma was open-minded about letting us pick the length of the trading period for calculating an average, then we might be able to bridge the gap between us.

"Let us do some research to see what range of days would be most favorable to us," I offered. "We would also look at some market precedents for pricing CBs, in addition to the Buffett/Goldman case."

I knew the market precedents would be useful to Ma, as he would need those to help convince his team, his board, and eventually his shareholders. His primary concern, when it came to the pricing for SDB stock and for PAIG's CB, remained market perception, given how battered PAIG's credibility had been by the debacle of its Fortis investment.

Regarding the idea of our making an additional investment in PAIG itself, Ma said he was no longer keen on it. PAIG did not need the cash, he said, and bundling two deals together would be too complicated for regulatory approvals. I understood his concern and suggested only that he keep the door open for us. We liked his franchise and we wanted to be his long-term partner. But we could consider all that after the SDB deal was done.

I sent an email to my partners that day. It concluded this way: "I think that, subject to diligence on [PAIG], what he has proposed is very close to what our ask is and if we can get a CB of [PAIG] with a conversion price at or near the market with or without a modest coupon, it is a deal we should consider."

Ma and I spoke again the next day, January 8. He indicated that a conversion price for PAIG's CB based on the lower of one-month or two-month averages was okay with him. The one-month average was about a 9% discount from PAIG's closing price of January 7, and the two-month average a discount of 20%, although the price fell another 6.5% on January 8. The bottom line was that he was willing to give us a discount for the CB conversion price. He added that he didn't want the coupon or interest rate on the CB to be more than 1%. "My offer to price your SDB stock at 18 [yuan per share] is premised on SDB's stock price staying above 10," he warned. "Or I can only pay you a premium no greater than 80%."

That was understandable. He had his constituencies to answer to, and he had to take care not to pay too high a premium. I told him that SDB would announce the write-off and the special provision for the NPLs in the following week. I didn't know what impact this would have on its stock price. I suggested we wait until after the announcement to take our conversation further.

Ma lamented the fact that PAIG had reached the 5% regulatory limit in its holdings in SDB stock. "Or we would continue to buy, and our buying would support SDB's price," he said. "In any case, let me get the ball rolling internally to build consensus first."

Ma might have been getting ready to do a deal, but my partners were now less than enthusiastic. In response to my report on these conversations, Dick Blum responded:

I think the key issue is the point that Shan brought up last, which is, should we even be thinking about selling the bank at this point? The shares are down 70 percent from what we thought we might be able to sell it for some months ago. My belief is that Frank [Newman] has done a very good job of restructuring the company and that with

*substantial deficit spending from the U.S., Europe, China, India and
even Bhutan, we are going to have more than $5 trillion worth of new
liquidity in circulation over the next year or two. I think after all that's
transpired here, the last thing we should do is to sell this institution
way too soon.*

Bonderman echoed Blum's view: "I saw Dick's memo with respect to
SDB. Assuming we actually have a choice here, Dick makes a good point
as to whether we should do this at all."

Ma called me on the evening of January 13. SDB's stock price had
actually held up quite well after its announcement of the massive write-
off and provisioning that day, and he had taken note. Now he wanted
to move forward. He reiterated a price of 18 yuan for our SDB stake
and suggested a conversion price for PAIG's CB equal to the average
price of PAIG stock over 10 to 30 trading days, whichever proved more
favorable for Newbridge. The coupon—the interest—on the CB would
be 1% per year.

I told him that we hadn't yet decided what to do. I would go to the
United States for our annual offsite meeting and speak with my partners
about his proposal. Ma asked me to meet with Louis Cheung, the PAIG
president, before I left to talk about the basic parameters of a deal.

★ ★ ★

Meanwhile there was another major development on the global stage,
one that hadn't been driven by the markets: As of January 20, 2009, the
United States had a new president. And I received a wonderful opportunity, thanks to my colleague Dick Blum—he provided me two tickets to
the inauguration of Barack Obama.

The one member of my family who could join me was my son Bo,
who was pursuing his master's degree at the Wharton School in Philadelphia. And so it was that I was able to experience this truly historic day
for America with my son. It was a clear morning in Washington, with
blue skies and bright sunshine over the nation's capital, but it was also
freezing cold. The long Washington Mall was packed with people. But
Blum's tickets were special: Bo and I found ourselves seated next to Oprah

Winfrey, Jesse Jackson, Denzel Washington, and some other celebrities. We weren't that far from the new president himself when he took the oath of office on the Capitol Hill balcony.

I didn't even realize Denzel Washington had been sitting next to us until I was back home and my wife pointed him out in the photos we had taken. With a full beard, he looked nothing like the man we knew from the movies. She chided me for missing a great opportunity.

"How can you not recognize him?!" she demanded, quite disappointed. She is a big fan of Denzel and his movies. I am, too. "Well, he didn't recognize me, either," I replied. She almost choked on her food.

I flew from Washington to San Francisco the next day; Bo went back to Philadelphia. On January 22, our partners met in TPG's office on California Avenue to discuss PAIG's latest proposal. Newman dialed in from China.

I described the deal in detail and made the case that as currently structured it was attractive and a good option for us. I explained that we would be able to capture both a sizable control premium and the upside built into the SDB investment when the market recovered; all the while we would have our downside protected if the market continued to dip because we would be holding a convertible bond of PAIG.

The logic here was that there was a strong correlation—not perfect, but significant enough—between the SDB and PAIG stocks, as both were financial institutions in China, and as such they usually rose and fell in tandem with one another. PAIG's convertible bond would increase in value if its stock price rose, allowing us to sell the CB to pocket the gain. If the market slid further and stayed down in a protracted global recession, we could redeem the bond at its maturity to receive the full principal value plus interest.

After two hours of deliberation, we reached consensus, and I had my marching orders. We were to proceed with PAIG.

While markets elsewhere continued to drift downward, the Chinese stock markets started to look up. By the time I met with Louis Cheung and Richard Jackson in Shenzhen on February 9, SDB's stock had jumped more than 40% since the first of the year. PAIG's stock was traded on two stock exchanges: Shanghai, referred to as the A share market, and Hong

Kong, or the H share market. Whereas PAIG's H shares had traded at a discount to its A shares, now those prices had converged. This made a bond convertible into its H shares more attractive; Hong Kong's stock market was far more liquid than Shanghai's.

My colleague Jessica Li and I spent the whole day in PAIG's Shenzhen office with Cheung and Jackson, but we finished the meetings without a deal. PAIG wanted to cap the price for our SDB shares at 18 yuan, regardless of the earlier proposed 80% premium. I felt that price would have made sense when SDB stock had traded at about 10 yuan. But now, SDB's stock was already at 13.75, and PAIG's stock was also trading significantly higher. If we agreed to a cap of 18 yuan, not only would the nominal premium for our stock be reduced by about 30%, but we would also have to pay a premium for PAIG stock. I argued that we should use the same methodology, including the agreed premium, to calculate the exchange ratio and keep both parties in the same economic position. A cap for SDB's stock price, I said, didn't make sense unless we also capped the conversion price for PAIG's bond.

Furthermore, Cheung told me that PAIG's advisors, now including the Chinese investment bank CICC in addition to Goldman Sachs, had advised them that there was little chance PAIG would receive the necessary regulatory approvals to issue a convertible bond. He asked me if we would still like to give it a try, suggesting PAIG would make an effort to obtain approvals if we wished. I said no; if PAIG and CICC both felt such an effort would be futile, there was no point in our pursuing it further. We would both be subject to the risk of either long delays or no approval.

Jackson suggested instead that we revisit the concept of PAIG giving us a put right for our SDB stake. I said we would be happy to take the put option if we combined it with the right for us to buy PAIG stock at an agreed price. Such a combination would, in my view, get us to the same place as a convertible bond—a put right would be equivalent to the right to receive payment in cash when a bond matures and the right to buy PAIG shares at a pre-agreed price would be the same as exercising the right to convert a bond into shares of the issuer. Both sides agreed that a put right combined with our right to buy PAIG shares would have a better chance of obtaining regulatory approval than a straightforward convertible bond issued by PAIG, even though the two things were equivalent in value and rights for Newbridge. Nonetheless, PAIG advisors on the CICC team wanted us to accept a share swap, exchanging our SDB shares for PAIG shares, without giving us the option to choose to be paid in cash—exercising our put right, or in PAIG shares. I rejected the CICC's idea.

Although we parted company without a deal, it seemed to me that both parties were motivated to find a way to make one happen. We agreed to keep trying.

On March 4, a large contingent of TPG/Newbridge partners including Dick Blum, Jim Coulter, Jonathan Coslet—TPG's CIO—Tim Dattels, and others descended on PAIG's Zhangjiang headquarters in Shanghai. I had coordinated the visit with Ma. He brought out his top lieutenants and cut short his attendance at a national congress meeting in Beijing to join us. This was my fourth visit in six months to the PAIG campus. I felt almost at home there.

That morning, our hosts took us on a tour of the campus. By now I could have given the tour myself. This was followed by a meeting in the boardroom with Ma, Cheung, Jackson, and others. Then we had lunch together. Ma again spoke passionately about his vision, his strategy, his operation, and his beliefs. I felt he was at his best, and I had no doubt he impressed his guests. We touched only briefly on the deal. He said he wanted to try the convertible bond structure again.

My partners came away impressed. That was not only Ma's purpose, it was mine as well. I wanted them to be as taken with Ma and PAIG as I had been. My partners would only be willing to support the deal if they liked PAIG and respected its leaders. Before that day, their views of PAIG had been based purely on internal memos and discussions, which, while positive, lacked any real, tangible qualities. The visit had made a difference. We were all convinced that PAIG was a quality company whose stock we would be happy to own.

But the visit also had an unintended consequence from my point of view. My partners had become too eager to cut a deal with Ma, and some had grown impatient with me as the negotiations dragged on. I believed PAIG would be a good partner; I just didn't believe the deal on the table was good enough yet.

I met with Ma a couple of more times in March, to try to hammer out a deal. But we could not make progress on pricing or the structure. By early March, SDB's stock had already climbed back to 15 yuan per share, but Ma was still insisting on a cap of 18. His team was also resisting the CB idea, citing the difficulty of getting regulatory approval. I had done my own homework to learn that while it wouldn't be easy, there were precedents for China's H share companies—those listed on Hong Kong's stock exchange—to issue convertible bonds. I remained convinced a CB was still our best bet.

The PAIG team was at least making an effort. After checking with their advisors and regulators, Cheung and Fred Hu of Goldman informed me that they had concluded that a CB was doable after all. However, Cheung said his side would only agree to the CB structure without fixing the exchange ratio, or a straight share swap without a CB. What I wanted was a CB *and* a fixed exchange ratio. We each understood what the other wanted, but continued to talk past each other.

Within our camp, some partners insisted that we needed to get a portion of our sale to PAIG in cash. I disagreed, for two reasons. First, a cash sale, even partial, would mean our immediate exit, which might not be acceptable to either the government of Shenzhen or the banking regulator. Second, we would have cashed out without the possibility of capturing any potential upside when the share prices of SDB or PAIG rebounded more strongly than they had, something I thought was inevitable. Another partner thought that SDB's price already contained what he called "acquisition fluff," which would dissipate if the deal with PAIG fell apart. Therefore, he was gently putting pressure on me to come quickly to an agreement with PAIG.

I was anxious as well, not so much because I was under pressure to sell our control in SDB, but because I knew that part of the plan—to issue new SDB shares to PAIG so it would be able to attain more than 50% shareholding—might be in jeopardy if the stock price rose too much. With the parties bogged down in negotiations, the market was getting away from PAIG. SDB stock continued its ascent, making the private placement and merger more expensive with every passing day.

With so much disagreement and second-guessing in our own house, I thought my hands were tied. Now that so many of our partners had met and gotten to know Ma and his senior staff, there were a number of parallel channels of communication at work, and different ideas were being thrown around. Mixed signals were being sent as to what we really wanted. I worried that these communications might undercut my bargaining power. With so many cooks in the kitchen, I decided to step aside to let others take over the negotiation, although I doubted that they would succeed where I had not. There were a lot of pieces to the puzzle that needed to be solved simultaneously.

I spent the next month preparing memos with my team, bringing my partners up to date on the various discussions we had had with PAIG, and building internal consensus on a term sheet to negotiate with PAIG. But I refrained from talking to Ma. I was with my Newbridge colleagues in San Francisco on April 14, where I printed out five documents for my partners—including a side-by-side comparison of various proposals, a couple of memos, a side letter, and a new term sheet.

The problem with long memos and supporting documents is that it's easy for the reader to get lost in the woods of detail, and to see trees but not the forest. Bonderman is known for his ability to cut through details, cut to the chase, and get to the bottom of key issues quickly. He reads voraciously and more rapidly than anyone I know, but he was probably frustrated by all our lengthy internal memos and presentations. He would ask, sometimes in exasperation, "What's the deal?"

To accommodate Bonderman's wish to see the big picture, TPG created a rule, no doubt at his request, that the first page—and one page only—of any deal memo or presentation for the IRC be referred to as the BUS, which stood for "Bonderman Ultimate Statement." The BUS would always provide a succinct description of a deal, a summary of key issues and points, pros and cons, and what decision was sought from the IRC.

When Bonderman asked me for a basic term sheet after seeing the stack of documents I had circulated, I knew exactly what he wanted. He wanted a BUS for this deal. On April 16, Bonderman sent an e-mail to all the partners. "At my request Shan put together a short 'basic term sheet,'" he wrote. "Comments please." The basic term sheet read:

a. *A convertible bond issued by PAIG which will be convertible into PAIG's H shares*

b. *3-year maturity*

c. *Redeemable by the investor at any time after 1st anniversary ["Redeemable" means we would have the right to ask PAIG to pay us the full amount in cash] d. Coupon paying a rate similar to the dividend payout rate for the H Common Stock of PAIG*

   e. The conversion price will be the one-week average of PAIG's H share stock prices at the time of the agreement

   f. Conditionality: None but [Newbridge] would make its best effort to cause SDB to issue new stock to PAIG through a private placement to the maximum as required by the capital needs of SDB and as permitted by regulation

   g. Pricing of SDB stocks owned by [Newbridge]: {50 percent} [brackets indicate subject to negotiation] premium to the one-week average of SDB's stock prices at the time of the agreement.

I suggested in the term sheet to lower the premium for our SDB shares from 80% to 50%, not so much due to PAIG's insistence, but more because of my own desire to find a compromise.

On April 20, we held an internal conference call to discuss the deal. After some discussion, all the partners agreed to my proposal. I was scheduled to meet with Ma in Shenzhen that day. We met over a box lunch in his office. Part of his plan, created with his Goldman team, was to take SDB private by an open offer to buy all the SDB shares. Although take-private transactions were common in overseas stock markets, I knew of no precedent in China, and so I was afraid it would render the whole deal too complicated to succeed. To ensure success, I wanted a deal to be as simple as possible.

I shared with Ma some of our experiences in the acquisition of SDB. I told him we had achieved a few breakthroughs. Buying control of a Chinese bank and owning legal person shares of an A-share listed company had been unprecedented and off-limits to a foreign financial investor. Consequently, our deal had required approvals at multiple levels of the government including the State Council. But we had no choice at the time; there was no shortcut to achieve our objectives. Had we not received the approvals, we would have had to give up and abandon the deal.

"But today," I said, "we both want the deal to go through without a hitch. Anything requiring special regulatory approval would create uncertainty in outcome and timing." To take a public company private was unprecedented in China. I could not even imagine how complicated the approval process would be because there were no established rules for such a transaction. "If we structure our transaction within the framework of established rules and regulations, we will know for certain we will be able to get our deal done."

Ma saw my point immediately. He was more familiar than I was with the difficulties in navigating the regulatory approval processes. "Great minds think alike," he said. "I was concerned about the same. There would be too much uncertainty if our deal required policy breakthroughs or represented precedents for regulators." He and I agreed then and there that we would avoid requiring policy breakthroughs or setting precedents in crafting our deal. We would endeavor to find the shortest of shortcuts to reach our goals.

★ ★ ★

By early May 2009, SDB's stock had risen to 17.90 yuan per share, up 89% from the beginning of the year and 22% from just two months before. PAIG's H share stock price had stayed flat. The market had swung definitively in our favor. If PAIG had to pay a large premium to meet our price expectations a couple of months earlier, now the price would be right for us with a much smaller premium to the market price. I could not sit on the sidelines any longer; I thought the time was ripe to cut a deal with PAIG. I called Ma to request a private meeting.

We met in his office on May 12. The deal structure we discussed was consistent with the "basic term sheet" I had prepared for Bonderman and that my partners had approved, with some tweaks to accommodate Ma's wishes. I sent him an email that evening to summarize and confirm what we had discussed. Among other things, I wrote: "The pricing of SDB shares should be the higher of 20 yuan per share and a premium in percentage terms to be agreed between the two sides."

To my surprise, Ma's response was rather subdued and noncommittal: "Thank you. We will send your proposal to my PAIG colleagues to study." With that sentence, he disavowed any agreement or understanding between us.

I wasn't sure where we stood. It took two to tango, and the dance could not be forced. I decided to sit tight for a while. To be patient, as we had been on so many occasions over the years. To wait and see.

Chapter 23

Turning Point

I n the end, I didn't have to wait long. The turning point arrived much sooner than I had imagined.

On May 14, two days after that last email exchange, Ma called as I was driving through a tunnel on my way to the office. I slowed down a bit to pick up his call.

"Weijian," he began, "are you talking with others?" He was referring to SDB, of course.

"Who told you?" I asked without answering his question. In fact, we weren't talking with anyone else, but it wouldn't hurt if he thought we were.

It turned out that a senior insurance regulator had told him that China Life and PICC, both major state-owned insurance companies and competitors of PAIG, were talking with Newbridge to buy SDB. He asked if I could put a stop to those negotiations.

It was the first time I heard that these two insurance companies were interested in buying control of SDB. But they had taken an approach different from PAIG's. Instead of talking with us first, they were checking with the regulator to see whether such a deal would receive its blessing. Knowing there was potential competition, Ma was spurred into action. Suddenly he wanted to get a deal done as quickly as possible.

I didn't respond to his request that we call off discussions with the others—there was nothing to call off—but I didn't have to tell him that. I said only that our preference was to do a deal with PAIG, but we couldn't just wait around. "How about if we both make an effort to reach a deal in the next few days," I suggested, "and we will make a final go or no-go decision by the end of next week?" That would mean a decision by May 22.

Peter Ma agreed.

I was convinced by then that without a deadline, a deal would never happen. Time and again we had learned a lesson in deal-making: There was always tomorrow, always a new wrinkle, and people would always want to squeeze more from the other side, using every bit of time they were given. We were still quite far apart; it would require much horse-trading to bridge the gap. I knew Ma had the authority and power to override his colleagues, and to act fast. Now he had to. On my side, I would only go back to the IRC once I thought I had a fully negotiated deal in hand. I was confident that if I liked the deal, my partners would like it too. I felt that the time had come to cut a final deal.

Nothing motivates a buyer more than the risk of losing to potential competition. Ma swung into action immediately. Less than an hour after his call, I found Fred Hu of Goldman, PAIG's advisors, along with his colleague Kenneth Wong and others, in our office. Over the next two hours, I analyzed for them why their proposal—for PAIG to do a general offer for all SDB shares using a combination of A-share-based convertible bond and cash—was not practical. It would carry too much risk, especially on the regulatory side. I laid out what I considered to be an easier path, involving three simple steps: PAIG would acquire our SDB shares, then it would buy another 20% of newly issued shares from the bank, and then SDB and Ping An Bank would merge through a share swap. I went through the basic terms I had discussed with Ma a few days before.

I had concluded that a convertible bond, which I had favored, entailed its own risks. There was no assurance that such a plan would receive the necessary approvals, or how long it might take to receive them. But it was possible to create a structure giving us the same rights and economic benefits as a CB without labeling it as such, and thus skirting the need for special regulatory approvals.

After the meeting, I pulled Hu aside. I thanked him and the PAIG team for having agreed to the CB structure, and for taking so much time and effort to research its feasibility. But I suggested that at this point the best way forward involved modifying the form of the structure, without altering its substance. To avoid the uncertainties that would come with a CB and the State Council approval process, I suggested PAIG should make Newbridge a binding offer to buy its SDB shares, in exchange for giving Newbridge an option to take either cash or PAIG shares. Such an option would essentially provide us the same rights and benefits as a CB, but it would not be a CB—and therefore would have no need for special approval.

Hu saw my point and agreed it was a workable idea. We jointly decided to go to Shenzhen to see Ma and his senior staff the following week.

For four days beginning on Sunday, May 17, I met or spoke with Hu and his colleagues every day in the hopes of settling all the details of a deal. At times, I thought we were getting close; other times, I was unsure. But soon a new deal structure was taking shape, largely along the lines I had proposed. PAIG would provide Newbridge with an irrevocable offer to buy Newbridge-owned SDB shares, and Newbridge would have the option to take either cash or PAIG shares. The "cash price" would reflect a premium of 60% over an average of SDB's stock price, although we hadn't yet determined the time frame for that average. The "share price" would be roughly equal to the cash price in value, but paid in PAIG H shares, whose number would be fixed at the time the deal was signed. The option would give us the right to take a fixed number of PAIG shares if they appreciated to a value greater than the cash price, or to take cash if PAIG's stock price fell. Thus, the minimum price we would eventually receive was the cash price, and we could get better value if PAIG shares were worth more.

On Tuesday evening, I got a call from Hu. He said the PAIG team had spent the whole day reviewing and discussing the various terms. I had proposed that we would have 24 months to exercise the cash-or-shares option. Hu said PAIG was willing to give us 12 months. I thought that was good enough. Then he said that PAIG had agreed to the exchange ratio of 1.65 SDB shares for one PAIG H share—the formula I had proposed for our option to take PAIG H shares for payment. I had suggested that the cash price be 10% less than the implied value of PAIG stock, based on the exchange ratio. Hu said Ma wanted a steeper discount.

I didn't budge. In my mind I was prepared to concede to a deeper discount, but I wanted to keep that option in reserve, in case I had to trade it with Ma for something else I really wanted. An exchange ratio of 1.65 would translate into a value of 28 yuan per share of SDB stock, compared with its market price of 17.89 yuan on that day—a premium of 56%.

The next morning, I was in the middle of a conference call when I noticed an incoming call from Hu. I dropped off the conference call to answer. Hu told me that PAIG could no longer accept the exchange ratio at 1.65 and was proposing to set a range of 22 to 26 yuan for pricing our SDB shares. I rejected this outright. It would have meant that the exchange ratio would not be fixed.

I thought I had to make an effort to break the logjam. To leave no ambiguity, I sent an email to Hu to explain my position and then slimmed down my proposal. PAIG was talking about a cash price range of 22–26 yuan for our SDB shares. I offered to take 22, the lower end of the range. I insisted on the exchange ratio of 1.65 SDB shares for 1 PAIG share, because I thought it far more likely that we would eventually take PAIG shares for payment than to take cash, in view of the upward trajectory of its stock. Already, the "share price" was about 28 yuan, about 27% higher than the "cash price" of 22 yuan. I hoped we would ultimately gain more from the appreciation of PAIG stock.

Hu called again just before dinner. He said they had considered my email and done some rethinking. Now he had a counterproposal, one that would require major concessions—such as capping the implied price for SDB shares at 26 yuan per share in computing the exchange ratio. After Hu's explanation, I indicated we could accept his changes.

With that, I thought we were done. Or at least, Ma and I had a deal. I had yet to tell anyone within our own organization about how this latest outline was taking shape. Now I sent an email to Bonderman, who was traveling in Europe, and asked him to call. Then I emailed Dan Carroll, to bring him up to speed. Although I had kept him informed along the way, I wanted to do another round with him as a sanity check. Was I missing anything? Were there any red flags I hadn't considered? Carroll was sharp and would quickly catch issues I might have missed. I also offered some color on the negotiations with PAIG:

> ... *I have had some very intensive discussions with the other side. I understand that there was so much emotion in their internal deliberations that someone even smashed a mug (I don't know why). In any case, I think we are getting very close, largely along the lines I briefed you but with some changes and we are in agreement on valuation/ swap ratio.... I think we are a day away, if all goes well, from agreeing on all major issues. Actually we are in agreement on all issues but I need to see them in writing to know for sure....*

Bonderman returned my call the next day, Thursday, May 21, while I was attending a board meeting at the Bank of China Tower. I came out to take his call.

"David, I think we got a deal," I began. "No more convertible bond. Just an option by us to take either cash or PAIG shares within 12 months." I explained the pricing mechanism and other details, pacing the hallway while I spoke. It took me just a minute or two to finish.

Bonderman was quiet, for perhaps three seconds. Three long seconds, it seemed.

"That works," he said. Then he added, blunt as ever, "Get it done."

And then he was off. Just like that, the call was over.

As I knew well by now, Bonderman was like this. Even for complex issues, he usually got it immediately and made decisions in rapid-fire fashion. I rarely needed to explain. And he wasted few words. I loved his style, loved it, as it fitted perfectly with me.

Just as I had gotten our top partner on board, the PAIG team came up with another issue. Hu called with a new set of questions. Given where the stocks of PAIG and SDB were trading on May 20, an exchange ratio of 1.65 would imply a premium for SDB of 59%. Louis Cheung now wanted to cap the implied premium at 60%, which meant the exchange ratio could be less than 1.65.

I would not accept the new condition. Instead, I offered to allow PAIG to walk away from the deal if the prices of the two stocks, on the eve of announcement, meant a greater than 60% premium for SDB. But

I would not agree to an automatic readjustment of the exchange ratio to keep the premium within 60%.

For 24 hours, Hu went back and forth between the two parties over this issue alone. He found no resolution. It was already Friday, May 22, the go-or-no-go deadline Ma and I had set. Hu and I agreed we would have to wrap up the discussion by day's end, one way or the other.

The clock ticked into the afternoon, and still no word from PAIG. At 4 p.m., I called Kenneth Wong, Hu's deputy on the Goldman team. Wong said they were still speaking with their client. They would get back to me as soon as possible. I had thought the remaining issue was a small one; apparently, it was major for their side.

It was about 5 p.m. when Hu called again to say that PAIG had accepted my proposal, with only a minor rephrasing: If the market should be such that PAIG's offering price implied a premium more than 60% for SDB shares on the day before the announcement, PAIG would have the right to renegotiate. They were also okay with giving us the right to walk if the implied premium was less than zero. I was fine with the word "renegotiate" as opposed to my original "walk." It made no difference to me because I was sure we would talk again if the deal fell apart at the last minute. But we were, in a manner of speaking, leaving things in the hands of God, or the markets, and we had to keep our fingers crossed.

Relief set in. With no outstanding issues left, I proceeded to draft a memorandum of understanding spelling out the details of the deal. I sent that MOU to the Goldman team at about 8 p.m. It noted:

> This MOU is not binding and the contemplated transactions herein are subject to definitive documentation and approvals. It is envisaged by the parties that definitive documents will be signed and submitted to the relevant Boards of Directors for approval prior to submission to shareholders and regulators for approval within one month from the date of this MOU. This MOU expires after one month from the date of signing, unless the parties mutually agree to extend it.

It was not until Sunday night, well past our self-imposed deadline, that I received a reply from Goldman. It came in the form of the revised MOU, marked up by the PAIG side. I was pleasantly surprised to see that PAIG had dropped the right to walk away over that 60% price premium.

And so there it was, after all that back and forth, all the angst and agony! It was certainly a welcome and significant change in their position. I knew how difficult this had been for PAIG, and I really appreciated their effort. I was also relieved, of course, knowing that now there was no chance the deal would be torpedoed by a freak day on the stock market.

After some intensive work, helped along by lawyers on both sides, the MOU was concluded in the office of DLA, the law firm representing PAIG, in Hong Kong on the night of May 27. Louis Cheung was to sign on behalf of PAIG; I would do the honors for Newbridge. But Cheung was delayed arriving from Shenzhen, and Liew Shan-Hock and I waited and waited for him to show up. Finally, I decided not to wait further as it was getting late. I signed the papers and had a photo taken with Liew to mark the occasion. And then I left for home, knowing that Cheung would put his signature on the papers later. It was a milestone for us all, even though the signing was somewhat unceremonious.

The author and Liew Shan-Hock (right), head of mergers and acquisitions of PAIG, after signing the term sheet sealing the deal for Newbridge to sell control of SDB to PAIG in the law firm DLA's office on May 27, 2009.

Peter Ma and I congratulated one another over the phone. We had come a long way and overcome many obstacles to get to this moment. We both knew there was still work ahead to finalize and close the transaction. No doubt there remained many problems to resolve. The good news was that we were on the same side now—we shared the common objective of getting the deal consummated. Ma was confident. "There are no problems that can't be solved," he said to me, "but there are people who can't solve problems."

That, I thought, was a problem-solver talking.

Over the next two weeks, the parties and their advisors worked on the final documents while PAIG conducted its due diligence on SDB. There was no need for us to do due diligence on PAIG immediately. The agreement gave us a year to decide whether we would take PAIG shares or cash. We could always do the work if and when we decided to take and hold PAIG shares. The beauty of the structure, in the end, lay in its relative simplicity.

Meanwhile, I flew to Beijing to visit with the central banker and the CBRC chairman, the two key regulators, to brief them on our proposed deal. The central banker simply noted my report but offered no comments. Chairman Liu of the CBRC told me that there was no barrier for insurance companies to own banks, and no ownership limit, either. I reported that part of the deal would involve a placement of new shares by SDB to PAIG, to raise capital for the bank. That was good news for the regulator, which always wanted banks to have more capital. I also called Chen Yingchun, the Shenzhen vice mayor. He said the city government would be happy to see the deal go through.

It was necessary for both parties to keep the pending deal secret before we were ready to make a formal announcement. We were still negotiating on the final documents—and as we knew better than most by now, no deal was final until it was final. Both SDB and PAIG were public companies, so any premature leak could disturb the stock prices and jeopardize the deal. But the circle of people in the know had widened considerably, with so many working on due diligence and other streams of work. It became increasingly tricky to keep the pending deal under wraps.

On Monday, June 8, Cheung called me at about 7:45 in the morning. He said a Chinese publication was reporting a rumor about our deal. I saw the report a few minutes later. The headline read "Breaking News: PAIG to Acquire SDB through Share Swap."

This was trouble. The headline alone was likely to have a big impact on the stock prices of both SDB and PAIG, and that of course could complicate our transaction. We decided to halt trading of both stocks. From experience I knew that our request to the exchanges had to be in by 9 a.m. With about an hour to spare, we scrambled to hand in a written request. It was doubly pressing for PAIG, because its stocks were traded on both the Shanghai and Hong Kong exchanges.

The Shenzhen Stock Exchange required a reason for trading suspensions. At this moment, the definitive documents had yet to be finalized. The announcement submitted by SDB read: "The reason for the suspension is that this bank will submit a capital raising plan to the board of directors for discussion, which may contain market sensitive information." This referred to the issuance of SDB shares through a private placement intended for PAIG but we stopped short of naming the party. PAIG's announcement suggested it was considering an investment in SDB. It wouldn't be hard for anyone who paid attention to put two and two together.

The trading of the stocks of both companies was suspended in time. The announcements touched off a fresh wave of media speculation about a possible deal between Newbridge and PAIG related to SDB.

The Newbridge and PAIG teams had been working on the definitive agreements and negotiating the details for about two weeks. At times the going got tough over technical issues, but the process was grinding forward, albeit slowly for my taste. Our team met with our PAIG counterparts and our respective lawyers at 9 p.m. on Thursday, June 11, in a last-ditch effort to close off the definitive agreements. We were racing against time; now that the news had begun to leak, we needed to move fast. We did not adjourn until about 2 a.m. We had resolved all the remaining issues and were waiting for lawyers to review the draft one last time.

★ ★ ★

As I have said in these pages, I tend to keep business and family matters separate. But given the nature of the work, the hours, and the travel, there have been times when the personal and professional mingled or collided. So it was in that last sprint to the finish line with PAIG; indeed, I had had an important family matter on my mind.

When I got home in the wee hours, my daughter LeeAnn was still up, and in something of a panic. Her high school graduation was to be the next day, and she had somehow lost her graduation gown. I comforted her until she fell asleep at about three in the morning. It turned out she had left her dress near the water fountain at her school. It was recovered, just in time for the ceremony. There was no way I was going to miss her big event. I managed two and a half hours of sleep and did my daily run before heading to Shenzhen to join SDB's board meeting. I would get the work done in Shenzhen and return in time for my daughter's graduation.

The main agenda item at the board meeting was to approve the issuance of new shares to PAIG through a private placement. I had expected the board approval to be smooth sailing because PAIG had offered to pay the full 20-day average price for SDB, as opposed to the 10% discount that was customary and permitted by the securities regulator. To my surprise, however, the directors raised many questions and the meeting lasted four and a half hours. In the end, the board unanimously approved the issuance, with Newbridge directors recusing ourselves. I felt such relief—this time for both professional and personal reasons.

I left the board meeting just before 3:00 p.m. to drive back to Hong Kong. On the way, I continued to work on the final documents and began to draft press releases, in both English and Chinese. I arrived home in time to go with my daughter and the rest of our family to her high school for the graduation ceremony. I was almost overcome with emotion when LeeAnn marched into the stadium with her class to receive her diploma. What a milestone in her life and in ours! There was no time for me to get some flowers for the occasion; I got a bag of bright red tomatoes from the grocery store by the gate of her school: They looked cheerful and celebratory enough to me. But my wife didn't let me present those.

After the ceremony, I returned home to work on various documents related to the SDB deal. At about 7:30 p.m. someone delivered the signature pages for the final agreements. I signed them. The exchange of these pages was handled by lawyers, again rather unceremoniously. But now the deal was sealed.

It was pure coincidence, of course, that my daughter's high school graduation took place on the same day we signed the final agreement to sell control of SDB. But it made *Friday, June 12, 2009*, a doubly significant and happy day for me. Our son Bo had arrived back in Hong Kong from Wharton, and the whole family went out to dinner to celebrate LeeAnn's graduation. What a great and exciting day!

Chapter 24

A Deal That Shakes

The day after my daughter's graduation, I sent Newbridge investors a memo on the finalized deal with this brief description:

> *NB has the right to elect to receive the consideration in cash or PAIG H shares. If cash, Newbridge will receive 22 yuan per share or 11.4 billion yuan [in total] ($1.68 billion). If shares, Newbridge will receive 299 million PAIG H shares immediately and freely tradable on the Hong Kong Stock Exchange. The number of shares is based on an exchange ratio of 1.74 determined at the time of announcement and it is calculated as the last 30-day average closing price of PAIG H shares divided by 26 yuan per share for SDB.*

It should be noted that although the implied SDB per-share price of 26 yuan at the time of the announcement might have seemed much lower than the price at its peak (40 yuan or so in 2007), the total value of our holdings came out to be about the same. Due to stock dividends and warrant exercises, our SDB shares had effectively split; we now owned many more shares than we had in 2007. Of course, we hoped to gain even more from the further appreciation of PAIG stock during the 18 months until the expiration of our option.

The saga of SDB was not yet over. The next phase required an intensive public relations campaign and lobbying with regulators for approval. Even though we had structured the deal so that each of its components complied with existing rules and regulations, taken as a whole the transaction was still unprecedented in China, and no single regulator could approve it. The deal would require the regulators of at least three industries—securities, insurance, and banking—to consent, and maybe even higher governmental authorities. That was in addition to the shareholders of both SDB and PAIG. How the market would react would have a direct bearing on the decisions of shareholders and regulators.

In other words, we still had work to do.

* * *

Despite the long standing on-again, off-again speculation about a Newbridge–PAIG deal, the news still surprised the market when it was announced. The story was picked up by virtually every major newspaper in and outside China. Generally, the coverage was positive, although some, especially in the foreign press, lamented the fact that Newbridge's pending exit from SDB marked the end of an era. I figured the media had had too much fun covering all the troubles we had gone through since 2002—going after the bank, taking it over, and trying ultimately to restore it to health—that perhaps the press was sorry to see us leave the stage.

The Economist echoed such sentiment in a report in its June 18, 2009 issue, entitled *Money from Another Time*. The subtitle was telling, as it suggested we were leaving too early: *A foreign banking expedition in China reaches a premature but successful end.* Clearly the newspaper's editors didn't think there would be another deal like this one. "The news, on June 12th, that [Newbridge] is to sell its stake to PAIG, a large Chinese insurer, draws another line under that era." It concluded the story by almost wishing we would fail to receive the necessary approvals:

> *The size of [Newbridge]'s returns has prompted reports that the deal may be blocked by officials. No private-equity firm likes having its exit strategy undermined but this would be no disaster. Barring another*

round of banking failures, no foreign firm will be granted a simi-
lar franchise in China for years to come. Even with strings attached,
[Newbridge] would be hard-pressed to find a more interesting place for
its money.

A positive reaction in the media was critically important for both Newbridge and PAIG, as shareholders and regulators alike were likely to be swayed by public opinions when making their own decisions. PAIG devoted significant resources to drive an effective PR campaign. I assisted whenever possible to help shape the message.

For the most part, the press coverage was positive—in some cases even ecstatic. The influential journal *Caijing*, an independent business magazine, devoted a 14-page cover story to our deal, titled "Ping An-SDB-Newbridge: A Deal That Shakes." *Caijing* declared the deal a winner for all three parties involved, and characterized it as a "perfect and beautiful exit" for Newbridge.

But what took us all by surprise was the publication of a commentary in the *People's Daily*. Entitled "Expecting Multiple Wins," the article strongly endorsed and supported the deal. As I told my Newbridge partners, it was extremely rare—in fact, totally unheard of—for the official mouthpiece of the central government to endorse a commercial transaction. The writer outlined the many benefits that the deal would bring to PAIG and SDB, including SDB's increased capital adequacy and access to PAIG's vast customer base. And it even spoke highly of Newbridge: "In exiting SDB, Newbridge will generate a net gain of more than 9 billion yuan," the article asserted—which was a substantial underestimate, but we did not mind at all. "If it elects to take PAIG shares, it may become a shareholder of PAIG and share the opportunity for further upside with the appreciation of PAIG value." Then it said, "It must be said that this is a cooperation which unites a strong one with a strong one, combines and complements advantages of the parties, creates a win-win-win and gives each of the three parties what it deserves."

We could not have set the tone better. It was exactly the message we had wanted to deliver to the market. I thought Peter Ma and his team must have been behind the *People's Daily* commentary in some fashion.

Ma told me no, they had been as pleasantly surprised as we were. But I added a caveat to the good news in a memo I sent to my partners:

> While it doesn't get any better than this in China, it doesn't mean approvals are assured. Indeed the commentary concludes: "We don't know at the moment if the market will change and the regulators will eventually approve this transaction. But we still loudly cheer for PAIG's 'ambition.' Perhaps it is PAIG's 'not being peaceful' (Shan's note: A pun: "Ping An" in PAIG means "peaceful and safe." Its antonym is "not peaceful" or "restless.") that let us time and again dream about the bright future of China's financial institutions, to see and experience the smoke of gunpowder in the battle of international financial markets. Even though, success may take time and failures may be many."

The last sentence was probably a subtle reference to PAIG's ill-fated investment in Fortis, which eventually had to be written off. In any case, the rave reviews from both the privately run *Caijing* and the government mouthpiece *People's Daily* delighted us and gave us more confidence that we would secure the necessary regulatory approvals.

More importantly, it seemed the shareholders of SDB were cheering the news as well. On June 15, the first day SDB stock resumed trading, its price jumped immediately by the daily limit of 10%. PAIG shareholders were enthusiastic, too. Someone conducted an online survey to gauge public reaction to the deal. In two days, there were 40,000 respondents, of whom 73% thought the deal was favorable to SDB and 70% considered it favorable to PAIG. Clearly most observers considered it a good deal for both companies.

★ ★ ★

On the Monday morning after the news broke, I drove to Shenzhen to meet with the senior management in SDB's head office.

This was no ordinary management team. We had brought them together and journeyed with them on the long road of transforming and building the bank. Now we were all proud of what we had accomplished.

I felt close to them. I explained the outlines of the deal, and the vision PAIG had shared with us for SDB and its future. I praised them for all they had done to rebuild the bank and its fortunes, and assured them that we were handing the bank into a pair of safe and good hands. Most important, I informed them that PAIG had agreed with us that for at least three years after closing the transaction, there would be no layoffs, no demotions, and no pay cuts, other than for cause. We were taking care of our SDB employees as we headed for the exit.

The officers in the meeting greeted the message with relief. I wouldn't say there were no lingering doubts, but I believed the promises made by PAIG had gone a long way to calm the nerves of employees and win their support.

Meanwhile, the sale of our stake to PAIG required approvals by its shareholders as well. PAIG, listed on both the Shanghai and Hong Kong exchanges, had shareholders in both places. Under the relevant rules, the two groups of shareholders had to vote separately, and a motion could only be carried if both groups approved. That required a long notice period, meticulous planning, and intensive lobbying. PAIG shareholders were still recovering from the shock of the loss from its Fortis investment. We figured they might not be predisposed to giving management the benefit of the doubt. Indeed, while SDB's stock hit the 10% daily ceiling immediately after the announcement, PAIG's H share actually *shed* more than 3% in value on the first day, followed by another dip of 5% the next day. Its shareholders, it seemed, were far from convinced the deal was a good one.

We had a two-pronged attack plan to win the necessary approvals. The first was an intensive PR campaign. The second involved speaking directly with shareholders and regulators.

The work with the media outside China consisted mostly of making our case, to the extent possible, by fielding questions from reporters who would write their own articles. In China, a well-known company like PAIG maintained excellent relations with all the major media outlets, as did SDB. The company could actively engage and influence the media—sometimes going so far as to help reporters write their articles and analyses of the company. Such self-drafted pieces were referred to as "soft articles," and often the reporters or publishers took the articles fed to them without much change or editing. I personally had to review, edit, and sometimes rewrite four or five of these soft articles on a daily basis, to make certain both facts and messages were correct before I gave final approval for their release. We were literally doing the job of the reporters for some of these publications.

On June 22, Ma and I met with the heads of six securities firms to discuss how they would go about helping us win votes from SDB shareholders. I urged them to make their best efforts to gather market intelligence and lobby shareholders. I impressed upon them how essential this work was; failure would be disastrous for all the parties concerned.

On June 29, SDB held an "extraordinary general meeting," or EGM, with shareholders voting either in person or via the internet. Shareholders representing 66% of all eligible shares cast ballots. It was probably the largest shareholder turnout in the history of SDB. The proposal was approved by 93.7% of those who had voted, and that was despite the fact that neither Newbridge nor PAIG (which already owned close to 5% of the shares) had participated. China Life was competing with PAIG to buy control of SDB at the last minute, after news of the PAIG deal broke, and it had rapidly accumulated a block of 95 million shares, representing about 6% of the total. We were resigned to the prospect that this block would vote against the proposal. In the end, it abstained. All told, "No" votes accounted for only 0.2%. This cleared a major hurdle in the approval process. The positive publicity, coupled with our lobbying efforts, helped secure the great numbers. No doubt it had also helped that we had put together a fundamentally good deal.

"That high approval rate," I told my partners, "should give regulators the political cover to approve the entire transaction."

PAIG's own EGM, scheduled for August 7, would be the next critical step, as it had to approve, or reject, the proposal of a specific mandate for PAIG to issue H shares to Newbridge if we chose to receive them. PAIG's management was confident, but I was holding my breath. When the day came, PAIG held two EGMs, one each for its A and H shareholders. When the tally came in, the mandate was approved by 99.98% of A shareholders and 98.97% of H shareholders, resulting in an overall approval rate of 99.74%. We could not have asked for a better outcome.

★ ★ ★

Peter Ma, chairman and CEO of PAIG, at the ceremony to mark the final agreement for PAIG to take over control of SDB from Newbridge on June 18, 2009.

The author and Louis Cheung (right), president of PAIG, exchanging signatures on the final agreement to transfer control of SDB from Newbridge to PAIG on June 18, 2009.

The regulatory approval process would prove to be more complicated than any of us had anticipated. The insurance regulator had to determine whether PAIG, primarily an insurance company, should be permitted to acquire a national bank and what sources of capital would be used for the acquisition. The banking regulator needed to evaluate the PAIG's qualifications as a prospective bank owner. And the securities regulator had the authority to either approve or reject PAIG's issuance of H shares in connection with our transaction.

Together with our respective advisors, the two parties divided responsibilities to communicate with each of the regulators, responding to their inquiries and addressing their concerns. It was not enough for the head of an agency to indicate support. It took months for working-level officials to evaluate every detail of the transaction and check all the boxes. The Chinese bureaucracy worked reasonably well within any particular agency, but we discovered that the agencies did not really talk with one another, and that there was no formal channel of interagency communication, making it almost impossible for them to act in concert. We sometimes found ourselves in the improbable position of acting as messengers between various commissions responsible for approving our own deal.

"Why don't you just pick up the phone and call your counterpart there to sort this out?" I asked a regulator at one point. He had refused a request from another agency to remove a sentence in their joint submission to the higher authorities. The sentence in question, in six simple Chinese characters, read "[the three regulators] formed a unanimous opinion" in the concluding paragraph of the draft submission to the State Council. The regulator who wanted the phrase removed had objected because his agency had gone through its own internal process, independent of the others, so he wouldn't agree to say "formed a unanimous opinion." The other regulator had argued that the higher authorities would never approve language that didn't make clear that all three had reached a consensus view. It was all a bit mind-numbing. Eventually I persuaded the second regulator to make a very minor change, removing a single comma, and a single character, so it read that they separately reached the same conclusion. And that took care of that problem.

Still, the process dragged on. Senior regulators at the three agencies had told us they approved the transaction. Yet, as I observed to my partners in a December 2 email, they "definitely don't like each other." I went on to say:

> Getting all these government agencies to agree to something and to approve this deal is so painfully complicated. I still hope that we can get it done in the first quarter [of 2010]. Peter Ma and his team are also quite focused, but it seems the bureaucracy cannot be hurried, and inter-agency communication is almost impossible.

It is customary for a contract to include a termination date, after which the deal is abandoned if it is still not closed. Lawyers call it the "drop-dead date." For our deal, that date was set at December 31, 2009, which gave us about six months to get all the approvals. But as the date drew near, it became clear it wasn't going to happen. Neither party wanted to "drop dead" prematurely, so we formally agreed to extend the date for another four months.

Under the deal, we had until December 24, 2010, to exercise our option to choose to be paid in cash or in PAIG shares. In mid-March 2010, the securities regulator told us there was no precedent for approving a company's issuing H shares but the company not doing so in the end. Therefore, the regulator needed confirmation as to whether Newbridge would definitely elect to receive payment in the form of PAIG's H shares before it could give PAIG the approval to issue such shares. We were forced to make a choice long before we had anticipated.

By now, it was already six months after the announcement. PAIG's H share was traded at HK$63. The cash price for our SDB stake was equivalent to HK$43 per PAIG share. In other words, the share price was now about 50% higher than the cash price. The H shares we were to receive from PAIG were immediately tradable, and given that PAIG had a market capitalization of some $60 billion by now, there would be enough liquidity for us to dispose of the shares in the market fairly quickly. In view of all this, I communicated to the regulator our decision to take PAIG shares.

It turned out that even with the extension of the drop-dead date, we hadn't left ourselves a single day to spare. It was not until the last day of April 2010 that PAIG finally received approval from the securities regulator, the last of the necessary approvals from the three regulatory agencies.

Now we could proceed to close the transaction.

On Monday May 3, 2010, we initiated the process of transferring our SDB shares to PAIG through the Shenzhen Stock Exchange. We received payment in 299 million PAIG H shares on May 6. And so, after all those meetings and calls and long nights and pestering regulators, the transaction was closed.

★ ★ ★

We sold the PAIG H shares in the Hong Kong market through a pair of block trades on May 13 and September 2, 2010, respectively. The term "block trades" refers to selling large quantities, or blocks, of shares of a publicly listed company through brokers, typically after the stock market closes. Brokers in turn sell those shares to large institutional investors. Through such trades, large volumes of shares can be sold at once. If a shareholder wanted to dribble its shares little by little into the market, it would take a very long time to dispose of such a large quantity.

The first trade of 160 million shares was priced at HK$60.60 per share, for a sum of $1.25 billion. The second batch of 139.1 million shares was sold at HK$64.99 per share, netting us $1.16 billion. In total, Newbridge received a payment of about $2.4 billion. After deducting a $139 million loan we had taken to exercise the warrants, the net proceeds of $2.27 billion represented more than 14 times our original $150 million investment in SDB.

By then, the global and Asian markets had rebounded strongly. In 2009, Hong Kong's Hang Seng Index rose 49%. PAIG's H shares spiked even more dramatically—by 75%. In early 2010, Hong Kong's stock market hit a lull and PAIG's H price also softened. By the time we sold our first batch on May 13, 2010, it had drifted down by about 10% from its peak at the end of 2009. The price had recovered by about 8% when we sold the second tranche on September 2, 2010. Overall, the value of PAIG shares we received was more than 40% higher than the cash price. The option (to take cash or PAIG shares) that we had structured in the sale of the SDB stake had worked out nicely for us.

★ ★ ★

While the timing and the structure of our exit were nearly perfect, there was no question that the main driver of our success in the SDB investment had been the successful transformation and build-up of the bank. And for that we owed much to the leadership of Frank Newman and his team. The strong growth of the Chinese economy had also put wind in our sails, as it had for many financial institutions. But SDB stood out among its peers. At the end of 2009, SDB was recognized with six best-in-class awards in an annual survey jointly conducted by the Chinese Academy of Social Sciences and other organizations. The awards included, most notably, "Best Retail Bank," "Best Brand in Bank Wealth Management," and "Most Respected Bank in China." We were especially proud of that last honor. I thought back to my first talk with the staff of the risk management department five years earlier, when I'd asked for a show of hands and the bank staff had rated SDB the worst among its peers. The bank had since then defied all the odds—having gone from worst to best in just five years.

Indeed, the SDB turnaround had exceeded all expectations, including our own. Between the end of 2004, when we first invested, and the end of 2010, the size of the bank in terms of total assets had nearly quadrupled, representing an annual growth rate of about 24%. Its net profit jumped by an astounding 20 times—or about 66% per year—from roughly $35 million in 2004 to just shy of $1 billion in 2010. The NPL ratio, that indicator of asset quality, which had worried us so much at the start, had dropped from a reported 11.4% in 2004 (the real number was a lot worse) to an actual 0.6% by the end of 2010. We had grown SDB's capital ratio from 2.3% (again, reported) to a solid 10.2%.

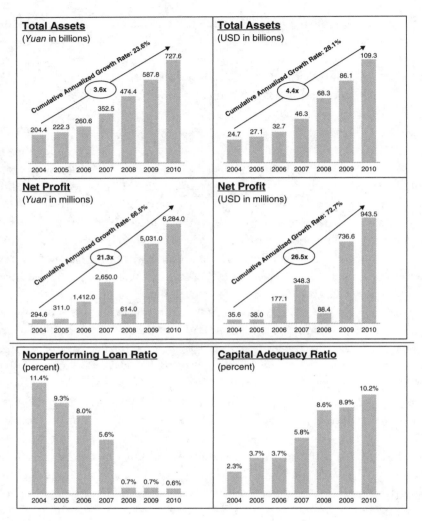

Transformation of Shenzhen Development Bank in key metrics
Source: SDB Annual Reports, 2004–2010.

★ ★ ★

Newbridge's investment in SDB was hailed as a landmark transaction. As I have frequently said, the whole endeavor of a foreign investor taking control of a national bank in China was unprecedented. And turning a deeply troubled bank into a healthy and strong one was a whole other story. The *Financial Times* used a quote from me in its report on the closing of our SDB sale on May 6, 2010: "We never regarded this as a simple investment but rather as a part of China's overall banking reform and we're very happy that after five years this bank has been fundamentally turned around from a weak bank into a very strong bank."

I have no intention of describing our objectives and experiences in overly lofty terms. After all, we were just investors, managing money for our limited partners worldwide, most of whom in turn managed money for pensioners, workers, and ordinary people. We had a fiduciary duty to increase value for their capital. That was our job. Over the years, we made some good investments and generated good returns for them. We also had our share of setbacks and failures. But our work brought together capital and talent to meet the needs of businesses. In the transformation of SDB, we did create value, not only for our investors but also for the institution itself and all those around it, not unlike the way a doctor treats patients, or a plumber fixes a leaky pipe, or a crew rescues a stricken ship. This, I believe, is how private equity makes the market more efficient as it seeks to maximize the value of its capital. Quite aside from our own intentions, the impact of our deal on the reform of China's banking system was quite profound and obvious in retrospect.

★ ★ ★

SDB was eventually renamed Ping An Bank, even though it was in substance the surviving entity of the merger between the two companies. It remains listed on the Shenzhen Stock Exchange under its original serial number, 000001.

It, and PAIG, continued to ride China's strong economic growth to new heights. In the 10 years between 2010 and 2020, the Chinese economy had grown, on average, 9.2% per year, its GDP more than doubling from less than $6 trillion in 2010 to $15 trillion in 2020.

When Newbridge had first acquired control of SDB at the end of 2004, its total assets were less than $25 billion. By the time we sold the bank in 2010, the total assets had more than quadrupled to $109 billion. As of this writing in 2021, the renamed Ping An Bank's total assets have exceeded $500 billion, about five times its size in 2010—and more than

20 times its size when I'd first laid eyes on this struggling bank in 2004. Who could have thought?

Right after PAIG took over control, Frank Newman stepped down as chairman and CEO, his historical mission accomplished to the satisfaction of all. Xiao Suining was made chairman and relinquished his position as president in a move that effectively kicked him upstairs. Richard Jackson became the bank's first president under PAIG control. Xiao retired a few years later, although we kept in touch. Jackson did not stay long, either. Hu Yuefei, the former Guangzhou branch manager I had challenged in 2005 to either accept a promotion or quit, became the president of Ping An Bank in 2016.

To mark the formal transition of power, a party was held to bid farewell to Newman and a few other senior members of management, and to welcome Xiao and Jackson as new leaders of the bank. Newbridge had acquired control of an ailing bank. And now, Newbridge was formally handing a healthy, strong, and much larger financial institution to PAIG.

Farewell party thrown by senior management members of SDB for Frank Newman on June 18, 2010, with Richard Jackson of PAIG. From left to right: Wang Bomin, Hu Yuefei, Frank Newman, Richard Jackson (PAIG), Xiao Suining, Liu Baorui, and Wang Ji.

PAIG, which had a market cap of about $15 billion when it had first approached us to buy SDB, was worth approximately $240 billion in 2021. Peter Ma remains its chairman. Ping An Bank, with SDB at its core, has helped transform PAIG into a diversified financial behemoth with the nationwide bank as its central pillar; its total assets reached about $1.5 trillion by the end of 2020.

★ ★ ★

I have metaphorically likened the SDB to a vessel: a leaky, sinking one to start out, and a totally refurbished one sailing at high speed in the end. When I reflect upon this journey, the image of barge haulers on China's Yangtze River comes to mind. I was struck, when I first saw them, by how hard their work was. I know from experience what hard labor is, but the job of a barge hauler was the hardest of hard labor. These men had to pull fully loaded barges and vessels upstream, hauling with all their might against strong currents, thick ropes wrapped around their shoulders at one end and tied to the barge on the other. They moved slowly, one step at a time, on the steep and rocky banks of the river. They were naked, save sometimes for a loincloth, and the sweat on their lean but strong bodies glittered under the scorching sun. As they pulled, their bodies bent so low that their hands touched the rocks under their feet. To synchronize their steps, they hummed—a rhythmic, melodic sound. They couldn't relax for a single moment, or the current would rip the barge away, and drag it down the river.

I felt our experience with SDB had been similar. Of course there was no physical pain, no back-breaking labor. But in our own way, we had pulled that barge upriver, against strong currents. Had we relaxed, it would have capsized and crashed. After more than five years, we had dragged our vessel upriver, navigated the shoals, and pulled it into a vast inland sea that was so huge that water stretched into the horizon. We didn't do it all by ourselves. We had a large and expert crew. It hadn't been a direct journey; there had been all manner of detours and setbacks, and so many moments when it seemed that the proverbial vessel would never reach its destination.

And now? Now we had disembarked, come ashore with our share of the cargo. Now, we hoped the new captain would take our vessel further, out into the open sea. We could imagine, in the distance, the sight of its full sail, a sail we had helped raise.

Epilogue

The Kingdom of Freedom

By January 2009, I had been at Newbridge for 11 years. The firm was called TPG Asia now—a rebranding that was the culmination of a series of events starting around 2006 in connection with the integration of Newbridge into TPG.

At the time of Newbridge's founding in 1994, there were no truly global private equity firms. And Asia was considered a backwater for institutional investors—not worth the risk. So Newbridge had been set up as a separate fund, independent from either Blum Capital or TPG, its sponsoring organizations, to accommodate those investors who were particularly interested in allocating some capital to Asia.

By the mid-2000s, global investors' perceptions of Asia had shifted. Newbridge, with its successful and high-profile deals, probably helped. Large U.S.-based private equity firms had expanded across borders, particularly into Europe, as their investors increasingly appreciated geographical diversity. What about Asia?

It would not make sense for TPG to set up its own operation in Asia to compete with Newbridge, given the affiliation between us. Therefore, the senior partners of TPG proposed integrating Newbridge into TPG. Essentially, Newbridge would become the Asian arm of TPG, and Newbridge's partners would become TPG partners, although generally Asia would invest out of Asian funds, not TPG global funds.

There were four senior partners at Newbridge: Dick Blum, David Bonderman, Dan Carroll, and myself. But since Bonderman also represented TPG, it was up to Blum, Carroll, and I to decide whether to accept TPG's proposal. Blum was lukewarm—he was upbeat about Asia and wanted to keep Newbridge an independent entity. But Carroll and I liked the idea of being part of a global firm, and we always thought of ourselves as part of the TPG family anyway. TPG was also hot, a high-flying firm widely admired on Wall Street for the high-profile deals it had done. Young talent flocked to it. Eventually, Blum went along.

In 2007, we changed the firm's name first to TPG Newbridge, and then to TPG Asia. The name Newbridge was discarded. I only realized later that we had thrown away something very valuable. In "low-trust" societies—a term coined by Francis Fukuyama in his book *Trust*, and which includes most of Asia—a strong, respected brand name differentiated a company from its competitors more so than in "high-trust" societies. It was only over time that I realized how much we had underappreciated the power of our brand: Even today, more than a decade later, I still meet strangers in business who would exclaim, after a brief introduction, "Ah, you are the *Newbridge* guy!"

The name was not the only change. The center of gravity in decision making shifted from Hong Kong to San Francisco, where TPG was headquartered. Dan Carroll had been my co-managing partner for Asia since I joined in 1998; now he was replaced by a partner from Europe.

In January 2009, TPG Asia held its annual offsite meeting at the Mandarin Oriental Hotel in New York City. There, I said to Bonderman that I wanted to take a sabbatical, and then retire. I told him I was finding it increasingly difficult to bear the bureaucracy of a large firm, as I was not much of an organizational man.

Bonderman pointed out that as a partner with some successful deals under my belt, I had earned the right to a fair amount of freedom at TPG. "You can do whatever you want," he said. "Why do you have to leave?"

"*You* can do whatever you want; you are the founder," I replied. "Who am I to do whatever I want?"

"No. You can do whatever you want," insisted Bonderman. He proposed to give me the title of "senior partner," a title nobody else held in the firm, and a dedicated team, so that I could do whatever I wanted, at least to his mind.

I was persuaded. The next day, as I was sitting together with Clive Bode, TPG's chief counsel, to document what Bonderman and I had agreed, he walked by.

"Retirement isn't an option," Bonderman said to me. "We will have to protect you from the bureaucracy." Then, he added: "Or protect the bureaucracy from you. I don't know which is more important."

How could you quit on someone like that? Besides, I knew I had to see through our exit from SDB. I stayed and tried to make myself harmless to the bureaucracy.

When I saw him again, in June of that year, I said to him, "David, I have removed the senior partner title from my name card."

"What are you now?" he asked me.

"I carry no title." Then I explained that I thought the title was pretentious and I didn't need it.

"Well, I am way ahead of you."

"What are you?" It was my turn to be surprised.

"I don't carry a name card," he said, coolly.

My wife and I traveled to Honolulu in August for a vacation with our daughter before she started college. One day, Bonderman called and said: "Shan, do you and Bin like water sports?" I didn't know what he meant. It turned out that he wanted to invite us to join him in the South Pacific on his private yacht later that year. We gladly accepted the invitation.

In late October, after TPG's annual investors' conference, I flew with Bonderman, his wife, and some friends of theirs by his private jet from Los Angeles to Port Moresby, the capital of Papua New Guinea, picking my wife up in the Australian city of Cairns on the way. We boarded Bonderman's boat in Port Moresby and set sail.

I had brought a book with me, *Marx's General: The Revolutionary Life of Friedrich Engels*. I had read a review in *The Economist* and was intrigued. Whereas there were numerous biographies of Karl Marx, I had never seen one of his closest, if not only, writing partner. Engels had penned most of the articles Marx published in English and, as a rather wealthy factory owner, supported Marx's family financially throughout his life.

I had read some of Engels' work during my youth when I was a laborer in China's Gobi Desert. He had postulated the concept that human society would eventually evolve from "the kingdom of necessities" to "the kingdom of freedom." The idea was that as long as man had to work for a living—for necessities—he was not free to do whatever he wanted. But in the future Utopian society that he and Marx had envisioned, the supply of material things would become overabundant; making a living would no longer be an issue, he believed, and man would be truly free. For some reason, despite their intellectual power, Engels and Marx failed to understand that such a society would never be attainable because resources, especially new things, would always be scarce, not simultaneously available to all people at all times.

I had read Marx and Engels before I knew English, at a time when most other books were banned. I had never understood what Engels meant, though, because the official Chinese translation of "the kingdom of necessities" was closer to "the kingdom of certainty" in meaning. I had always been puzzled by what that meant—breaking free from certainty seemed an oxymoron. Now I realized that the Chinese translation was completely wrong, and actually the opposite of what Engels meant. "The kingdom of necessities" was really a kingdom of uncertainties, which was why man had to work for a living and therefore was not completely free to do whatever he wanted.

It also dawned on me then and there, on Bonderman's luxury ocean-going yacht, that I was personally in a financial position to move from the kingdom of necessities to the kingdom of freedom—free to do whatever I desired.

It was time to leave TPG.

A strong wind was blowing over the Pacific Ocean that afternoon and the water was choppy. But the yacht had stabilizers extending out from under its hull, and it was so steady that I could still run on the treadmill on the upper deck. Bonderman asked if anyone would like to join him in the smaller tender boat for some deep-sea fishing. I was the only volunteer. The tender cut through the swelling waves, bumping up and down, with a few fishing lines trailing behind. I was sure we were the only people fishing for hundreds of miles in every direction, yet, after an hour, we had caught a total of one small swordfish. Nevertheless, we had a thrilling time. I was glad to find this rare moment alone with Bonderman.

At one point, I plucked up some courage and said, "David, I've made up my mind to leave." I meant to leave TPG, not to leave this world by jumping into the rough sea. I knew it was an awkward time to bring up the subject; I was his guest on his boat for a holiday. But I didn't know if there was any better time.

Sharp as he is, he must have sensed that I couldn't be persuaded to change my mind this time. He simply said, "Well, we should keep a relationship."

That was it. By then, we had sealed the deal to sell SDB to PAIG, although we were still waiting for final regulatory approvals. Bonderman and I further talked after the New Year and we mutually agreed I would officially leave TPG on June 30, 2010, after the company offsite meeting in Barcelona, Spain. By then, TPG would have found my replacement and there would be a seamless transition.

Usually, at least in the finance industry, people quit their jobs at the beginning of the year, after collecting their bonus. The idea hadn't even crossed my mind when I spoke with Bonderman on that November day in the South Pacific. But he was always fair. In the end, the firm paid my full bonus and profit share, and even let me keep some of the shares in TPG, which had been granted after I had given my notice.

My plan was to raise a new fund of my own with a new team. I had agreed with Bonderman not to poach any people from TPG. There was no question that my new fund would compete with TPG Asia, doing buyouts and investments similar to the ones I had learned how to do from Bonderman and other partners of mine over the years. Despite that, Bonderman helped me where he could. About a month before my departure, a reporter at the *Financial Times* somehow heard about it and contacted Bonderman for comment. He usually shunned the limelight and rarely gave interviews. But he agreed to comment this time, on the condition that the reporter would give the firm a couple of days to inform its limited partners—we wanted them to hear about this from us first, not from the press.

On June 2, 2010, the *Financial Times* printed an article with the headline "TPG dealmaker quits to launch Asia fund." It reported my imminent departure and my plan to "set up a fund focused on Asia-wide opportunities." It quoted Bonderman as saying: "Shan is a smart and persistent investor and TPG looks forward to continuing our relationship with him in his future ventures."

What I had in mind was to raise a new private equity fund to do what I had been doing at Newbridge and TPG and, over time, build a platform that would include other "alternative asset classes" such as real estate investments, credit funds, hedge funds, and so forth. Private equity was a lumpy business—while the payoffs could be substantial, it might take years for an individual investment to bear fruit. Diversification would smooth out the cash flow and make the platform more valuable—it is always easier to value a business with a stable cash flow than one with sporadic gains or losses.

The press coverage was not a moment too soon. It caught the attention of potential investors, which helped my fundraising even before I started, as well as those from unexpected quarters.

Tony Miller was an old friend. I got to know him when he was a partner at Carlyle, a major U.S.-based private equity firm, and I was at Newbridge busy working on the KFB deal, shuffling between Hong Kong and Seoul around 1999. Now in 2010, he was a partner at a firm named Pacific Alliance Group.

After he heard I was leaving TPG, he met with me for lunch on his next visit to Hong Kong from his home in Japan. He inquired about my plans and suggested that I meet with Chris Gradel, the founder of Pacific Alliance, to explore whether there were areas in which we could cooperate in the future. I was glad to.

Chris Gradel looked even younger than his 39 years of age. Born to German parents, he had grown up in Northern Ireland during the Troubles, those decades of violence when the Irish Republican Army was still active. He went on to study engineering and economics at Oxford University. After graduation, he worked for the family business of Robert Pritzker, a Chicago-based tycoon, and was at one time posted to a conveyor belt factory in Tangshan, about 110 miles (180 kilometers) east of Beijing. It was a real experience for the young European. He became a consultant with the venerated McKinsey & Co., but three years later, in 2002, he left to start his own firm with $10 million in seed capital.

By the time we sat down at Cipriani, on the 12th floor of the old, Art Deco Bank of China building, Gradel was managing almost $3 billion in funds. Pacific Alliance Group was a successful investor in both public and private markets in Asia.

Gradel quickly came to the point: He proposed that we merge our operations. I was surprised.

"But I have nothing to merge with you," I said, pointing out the obvious. I had yet to leave TPG. "Oh," he said. "You will raise your big fund." I thought he was more confident in me than I was. I was reasonably confident, but I never count my chickens before they hatch. Now he was asking me to trade my chicks even before I had any eggs.

I was, however, taken by the idea. It would be a shortcut to achieve my goal of helping build a diversified platform, and I could tell that Gradel, young, energetic, full of vision, and yet practical, would be an excellent partner. In many ways he reminded me of Dan Carroll, with whom I had happily co-managed Newbridge for more than a decade.

At the time, Gradel's firm had acquired a significant interest in a Tokyo-based real estate investment firm by the name of Secured Capital Japan. Secured Capital had been co-founded in 1997 by an American, Jon-Paul Toppino. Toppino had been in this trade for more than two decades, investing in real assets in the United States, Europe, and Asia. Secured Capital had been so successful that it went public on the Tokyo Stock Exchange in 2004; at its peak it had a market capitalization of more than $1 billion. If I agreed to join forces with Gradel and Toppino, there was a unique opportunity to create a diversified investment platform that would encompass three major strategies: credit & markets, real assets, and private equity. No such platform had existed in Asia. It was an exciting prospect.

Gradel and I quickly reached agreement. At my suggestion, the platform would be renamed PAG. I would become its chairman and CEO. I would focus on raising a buyout fund as I had planned. When the announcement was made, Bonderman was initially upset, as he thought I had joined another firm. I told him no, it was a merger—the best deal I'd ever done.

A few months later, with the help of a new team, I produced a private placement memorandum (PPM) for the purpose of raising a $2.5 billion buyout fund. The PPM provided information about the new team, strategy, and the track record of principal partners. Since my track record in private equity had been entirely with Newbridge and TPG, I wouldn't release it without Bonderman's seal of approval.

When I next saw him in September 2010 in the Shangri-La Hotel in Singapore, he and Clive Bode, TPG's general counsel, had reviewed the PPM. He told me, quite simply, he had no problem with it, and he liked the fact that "you give credit where credit is due." I had not embellished my track record, or taken credit for any Newbridge deals I had not led. And I did give credit to others who had been involved. Bonderman appreciated it.

By the end of 2011, PAG's maiden buyout fund was on its way to achieve the $2.5 billion target. The full integration with Secured Capital was also completed. Even though I was chairman and CEO, I focused all my attention and energy on our new private equity business. Chris Gradel and J. P. Toppino led the credit & markets and real assets businesses, respectively. I knew we would make a great team.

Acknowledgments

This is my third book. Many people ask me how I find time to write, between my job as the executive chairman of a leading Asia-focused investment firm and other responsibilities. The answer is simple: writing a book requires no deadline, so I do it in my leisure time, and for as long as it takes. I thought of writing this book more than a decade ago and I began to type it on my laptop about five years ago. I didn't know how long it would take to finish it and I was not in a hurry.

My job, however, can never wait. Frequently there are deadlines to meet. After all, I am in the business of making deals and investments. When there is a live deal, I devote all my time and attention to it, day and night, weekdays or weekends. To be competitive, a dealmaker must be able to seize opportunities and move fast.

One episode of this book describes how a counterparty of ours missed the window of a great opportunity by less than an hour, and the deal slipped away from their fingertips, eventually costing them more than a billion U.S. dollars. This has happened to me and my partners as well, to our chagrin and regret. Often one cannot control the speed of decision-making processes. The key to success in dealmaking is best summarized in the Latin phrase *carpe diem*—seize the day.

So for me, work always comes first and it can never wait.

But committing to writing a book, and in my case three books, does take away all my spare time and requires strict discipline, sacrificing the time I could have spent with my family and friends. There is no time to socialize or to have fun. I apologize to the people around me for being quite boring for so long while I write my books. I am grateful to my family for their support.

I owe my deep gratitude to Bill Falloon, the executive editor at Wiley, my publisher. He has guided me through the publication of all three books, providing invaluable ideas and suggestions. I feel lucky to have found him, and I hope he at least doesn't feel unlucky to have stuck with me.

Tom Nagorski was my first editor, until his new job made it impossible for him to continue. He made me ever-conscious of writing in such a way that a layperson to banking could easily follow the storyline.

Tim Morrison edited the whole manuscript. I remember one thumbnail-sized cartoon in the "Pepper & Salt" section of the *Wall Street Journal* from many years ago. It shows one man talking to another across an executive desk, and the caption reads: "Please turn my scrambled words into prose." I think Tim did just that.

My son, Bo Shan, helped me quicken the pace of the story by suggesting cuts to any detours that he thought were not directly relevant to the main flow. His editing was critical and brutal, as he didn't have to worry about the sensitivities of the author, but was ultimately useful.

I very much appreciate the work of Christina Verigan as my editor. She came highly recommended by Bill at Wiley and she proved truly professional. She helped me with my previous book, *Money Games*, as well as this one, and gave me feedback from the critical perspective of a reader. Addressing her questions improved the story.

Rachel Kwok provided me with secretarial support. She saved so much of my time and helped me keep track of what I needed. I am immensely grateful to her for her help.

I thank Mssrs. Liew Shan Hock, Hu Yuefei, Zhou Li, and others for providing me with historical photographs that are exhibited in this book.

I also thank Purvi Patel and other members of the Wiley team whose work ensured the timely publication of this book and a quality product.

Weijian Shan
September 30, 2022

About the Author

Weijian Shan is co-founder and executive chairman of PAG, a leading private equity firm in Asia. Formerly, he was a partner at the private equity firm TPG and co-managing partner of TPG-Asia (formerly known as Newbridge Capital). Prior to his career in private equity, Shan was a managing director at J.P. Morgan and a professor at the Wharton School of the University of Pennsylvania. He holds an MA and a PhD from the University of California–Berkeley and an MBA from the University of San Francisco. Shan is a member of the Board of Trustees of the British Museum and a member of the International Advisory Council of the Hong Kong Stock Exchange. He is also the author of *Out of the Gobi: My Story of China and America* and *Money Games: The Inside Story of How American Dealmakers Saved Korea's Most Iconic Bank.*

Index